Environmental and Architectural Acoustics

Second edition

This revised new edition of a classic guide adopts a multi-disciplinary approach to the challenge of creating an acceptable acoustic environment. It draws on the same basic principles to cover both outdoor and indoor space. Starting with the fundamentals of sound waves and hearing, it goes on to the measurement of noise and vibration, room acoustics, sound absorption, airborne sound insulation and noise and vibration control. Each chapter finishes with useful problem-solving exercises to assist comprehension. The strong focus on techniques, and methods and standards, and with its lead into further more specialised material, makes this book highly useful for advanced students and professional engineers.

The new updated coverage of international standards makes this book even more useful. It also serves as a foundation reference for students of architecture and environmental engineering, including those new to the study of acoustics.

Z. Maekawa is Professor Emeritus at the Environmental Acoustics Laboratory in Osaka, Japan. He is a former Vice-President of INCE/Japan and past member of the International Commission on Acoustics.

J. H. Rindel is former Professor at the Technical University of Denmark and Managing Director of Odeon Room Acoustics Software, Denmark.

P. Lord is former Professor at the University of Salford, UK.

Also from Spon Press

Fundamentals of Medical Ultrasonics
Michiel Postema
978–0–415–56353–6 (hbk)
978–0–203–86350–3 (ebk)

Auditorium Acoustics and Architectural Design, 2nd ed
Michael Barron
978–0–419–24510–0 (hbk)
978–0–203–87422–6 (ebk)

Room Acoustics, 4th ed
Heinrich Kuttruff
978–0–415–48021–5 (hbk)
978–0–203–87637–4 (ebk)

Environmental Noise Barriers, 2nd ed
Benz Kotzen and Colin English
978–0–415–43708–0 (hbk)
978–0–203–93138–7 (ebk)

Building Acoustics
Tor Erik Vigran
978–0–415–42853–8 (hbk)
978–0–203–93131–8 (ebk)

Predicting Outdoor Sound
Keith Attenborough, Kai Ming Li and Kirill Horoshenkov
978–0–419–23510–1 (hbk)
978–0–203–08873–9 (ebk)

Environmental and Architectural Acoustics

Second edition

Z. Maekawa, J. H. Rindel and P. Lord

CRC Press
Taylor & Francis Group
Boca Raton London New York

CRC Press is an imprint of the
Taylor & Francis Group, an **informa** business

A SPON PRESS BOOK

Published 2019 by CRC Press
Taylor & Francis Group
6000 Broken Sound Parkway NW, Suite 300
Boca Raton, FL 33487-2742

First issued in paperback 2019

Typeset in Sabon by Integra Software Services Pvt. Ltd, Pondicherry, India

ISBN-13: 978-0-415-44900-7 (hbk)
ISBN-13: 978-0-367-86546-7 (pbk)

British Library Cataloguing in Publication Data
A catalogue record for this book is available from the British Library

Library of Congress Cataloging in Publication Data
Maekawa, Z.
Environmental and architectural acoustics / Z. Maekawa,
J. H. Rindel, and P. Lord. -- 2nd ed.
p. cm.
Includes bibliographical references and index.
1. Soundproofing. 2. Vibration. 3. Architectural acoustics. I. Rindel,
J. H. (Jens Holger) II. Lord, P. (Peter) III. Title.
TH1725.M34 2011
693.8'34--dc22
2010011589

Visit the Taylor & Francis Web site at
http://www.taylorandfrancis.com

and the CRC Press Web site at
http://www.crcpress.com

Contents

Preface

Since the first publication of this book 15 years have elapsed, and during this period there have been many developments in acoustic technology, mainly through the brilliant progress of the digital computer. Therefore, I welcomed the suggestion from the publisher to write the second edition, which includes these developments. However, the physical theory of acoustics has not changed. Though digital technology is an excellent research tool, it is not covered to any depth in this book. The purpose of this book, as mentioned in the preface of the previous edition, also has not changed, i.e. it is to provide a good understanding of the foundation of acoustics to all readers.

I took great pleasure from the participation of a new and excellent co-author, Professor J. H. Rindel, who was interested in the first edition, and sent me many kind comments for its improvement. I believe that the topics are covered in more depth and breadth than the previous edition.

I am indebted again to my younger colleagues, Professor M. Morimoto and Associate Professor K. Sakagami at Kobe University, with whom I had many useful discussions. I found their articles, including the results of their own research, helpful.

Sincere thanks are again due to the co-author, Professor P. Lord, for his kind critical reviewing and correct English.

All illustrations in this version, except those with some note, are reproduced by permission of Kyouritu-Shuppan Co. Ltd.

Z. Maekawa
Osaka, JAPAN

Preface

Since the first publication of this book 15 years have elapsed, and during this period there have been many developments in acoustic technology, mainly through the brilliant progress of the digital computer. Therefore, I welcomed the suggestion from the publisher to write the second edition, which includes these developments. However, the physical theory of acoustics has not changed. Though digital technology is an excellent research tool, it is not covered to any depth in this book. The purpose of this book, as mentioned in the preface of the previous edition, also has not changed, i.e. it is to provide a good understanding of the foundation of acoustics to all readers.

I took great pleasure from the participation of a new and excellent co-author, Professor J. H. Rindel, who was interested in the first edition, and sent me many kind comments for its improvement. I believe that the topics are covered in more depth and breadth than the previous edition.

I am indebted again to my younger colleagues, Professor M. Morimoto and Associate Professor K. Sakagami at Kobe University, with whom I had many useful discussions. I found their articles, including the results of their own research, helpful.

Sincere thanks are again due to the co-author, Professor P. Lord, for his kind critical reviewing and correct English.

All illustrations in this version, except those with some note, are reproduced by permission of Kyouritu-Shuppan Co. Ltd.

Z. Maekawa
Osaka, JAPAN

1 Fundamentals of sound waves and hearing

Sound can be visualised physically as a wave motion, which is transmitted through a whole range of elastic media. It is called a sound wave. On the other hand, it is also a sensation subjectively perceived by the ear, which is stimulated by the sound wave. This is referred to as auditory sensation, a phenomenon which is the subject of advanced research and comes under the general heading of psychophysiology.

1.1 Sound waves

A sound wave is transmitted through a medium that has both inertia and elasticity. The space in which sound waves travel is called the **sound field**. In a sound field the medium particles exhibit a repetitive movement backwards and forwards about their original position. Since a particle in the medium causes a neighbouring particle to be displaced by ξ, the repetitive movement produces a wave motion, i.e. vibration that is transmitted from particle to particle successively in the medium. The direction of the particle's movement is the same as that of the transmission path of the sound wave. Therefore, it is called a **longitudinal wave**. As shown in Figure 1.1, the medium particles are crowded together at a certain point, producing a high pressure while at a neighbouring point they are dispersed resulting in a reduced pressure. Two such points of condensation and rarefaction exist alternately in the wave motion. Thus, at a fixed point, the dense and rare parts of the wave arrive alternately and the pressure consequently repeatedly rises and falls. This pressure fluctuation is called **sound pressure** p and the velocity of motion of the particles of the medium is called the **particle velocity** v.

The number of fluctuations in 1 s is called the **frequency** generally expressed by f, the unit of which is the **hertz (Hz)**. The distance that a sound travels in 1 s is called the **sound speed** and generally denoted by c (m s^{-1}). If we let λ represent the **wavelength**, then

$$\lambda = c/f \, (m) \tag{1.1}$$

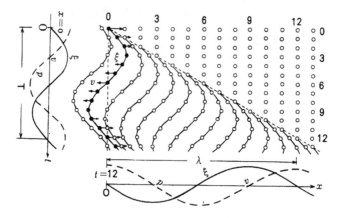

Figure 1.1 Particle movement and wave propagation; ξ: particle displacement, v: particle velocity; p: sound pressure; λ: wavelength; and T: period.

In order to express the wave motion in the form of a mathematical equation, the displacement ξ of a medium particle in the x direction is expressed as follows,

$$\xi = AF(t - x/c) \tag{1.2}$$

where F is a function which has two independent variables, time t and distance x; c is a constant and A is the amplitude, which is constant in this case. When a sound source produces simple harmonic motion the equation becomes

$$\xi = A \cos \omega(t - x/c) = A \cos(\omega t - kx) \tag{1.3}$$

where ω is the **angular frequency**; $\omega = 2\pi f$ and $k = \omega/c$. The sound pressure and the particle velocity can also be expressed in the same form. In Equation (1.3) when t becomes $(t + 1\,\text{s})$ and x becomes $(x + c \times 1\,\text{s})$,

$$\xi = A \cos \omega[(t + 1) - (x + c)/c] = A \cos \omega(t - x/c)$$

Therefore, the expression does not change. This means the phenomenon shifts through a distance c in $1\,\text{s}$, therefore, c is the sound speed. If we consider a certain position, for instance, where $x = 0$ in Equation (1.3), then $\xi = A \cos \omega t$, which means that it is a sine wave motion. Also at a certain time, for example when $t = 0$, $\xi = A \cos(-kx)$, which shows that the positions of moving particles also follow a sinusoidal pattern. In this case $kx = 2\pi(x/\lambda)$, therefore, when (x/λ) is an integer the values of ξ are the same, and it is found that particles separated by λ are in the same phase. Also, $k = \omega/c = 2\pi/\lambda$ and is called **wavelength constant** or **wave number**.

In the mathematical description of wave motion, use is made of complex number representation. For instance, the sound pressure p can be expressed as:

$$p = Pe^{j\omega(t-x/c)} = Pe^{j(\omega t - kx)} \tag{1.4}$$

where P is the pressure amplitude and $j = \sqrt{-1}$. Using Euler's formula

$$p = P[\cos(\omega t - kx) + j\sin(\omega t - kx)]$$

Therefore, if the real part is taken, the relationship has the same form as Equation (1.3), which represents a physical phenomenon while the imaginary part can be considered as something additional for calculation convenience.

The sound wave propagates in a spherical shape from a sound source whose dimension is relatively small compared to the wavelength. It is called a **spherical wave** and the sound source is referred to as a **point source**. Although an actual sound source has a finite size, it can be considered as a point source when the source is located from an observer at a sufficiently large distance compared to its size. When the sound source is located very far away, then the **wave front** approximates to a plane and the sound may be treated as a **plane wave**. Strictly speaking, a plane wave propagates only in one direction, and in a plane perpendicular to this direction its sound pressure and particle velocity are uniform and have the same phase.

1.2 Speed of sound

The sound speed c in a fluid whose density is ρ and has a volume elasticity κ is given by

$$c = \sqrt{\frac{\kappa}{\rho}} \ (\text{m s}^{-1}) \tag{1.5}$$

In the case of a gas, the pressure variations associated with the sound wave are adiabatic, therefore, when the ratio of specific heats under conditions of constant pressure and constant volume is γ (see Section 11.1).

$$\kappa = \gamma P_0 \quad \text{where } P_0 \text{ is the atmospheric pressure} \tag{1.6}$$

In air at $0°C$ and 1 atm of pressure, $\gamma = 1.41$, $P = 101\,300\,\text{N m}^{-2}$ (or Pa) = 1013 mbar, $\rho = 1.29\,\text{kg m}^{-3}$ and $c = 331.5\,\text{m s}^{-1}$.

Since ρ varies with temperature while the atmospheric pressure remains substantially constant, at $t°C$ the sound speed is

$$c = 331.5\left(1 + \frac{t}{273}\right)^{1/2} \doteq 331.5 + 0.61t \quad (\text{m s}^{-1}) \tag{1.7}$$

Table 1.1 Sound speed and characteristic impedance of various materials

Material	Sound speed c (m s^{-1})	Density ρ (kg m^{-3})	Characteristic impedance ρc (kg m^{-2} s^{-1})
Air (one atmospheric pressure, 20°C)	343.5	1.205	415
Water	1460	1000	146×10^4
Rubber	35–230	1010–1250	$3.5–28 \times 10^4$
Cork	480	240	12×10^4
Timber (pine, cypress)	3300	400–700	$1.3–2.3 \times 10^6$
Iron	5000	7800	39.0×10^6
Concrete	3500–5000	2000–2600	$7–13 \times 10^6$
Glass	4000–5000	2500–5000	$10–25 \times 10^6$
Marble	3800	2600	9.9×10^6
Sand	1400–2600	1600	$2.3–4.2 \times 10^6$

Reproduced by permission of Kyouritsu-Shuppan Co. Ltd.

Therefore, $340\,\mathrm{m\,s^{-1}}$ is generally used for calculation at normal temperatures. Also, the effect due to humidity is negligible.

The sound speed of a longitudinal wave in a solid, whose density is ρ and Young's modulus E, is expressed in the same form as follows:

$$c = \sqrt{\frac{E}{\rho}} \tag{1.8}$$

As shown in Table 1.1, the sound speed in a solid is much larger than in air except in the case of rubber, which is used as a special building material.

Equation (1.8) can be used to find the Young's modulus E of a material, from the measured value of c under suitable conditions. In a solid, not only longitudinal waves but transverse waves are also produced, therefore, every part of a building structure is subjected to complex vibrations (see Chapter 7).

1.3   Impedance

Generally, when some effect is produced by an alternating action at a point, Action/Effect $= Z$, called the **impedance** at the point. For instance, in an electric circuit where an alternating action due to an electric voltage E produces an alternating electric current I,

$$E/I = Z_e \tag{1.9a}$$

which is the **electrical impedance** at the point in the circuit. If E is a steady voltage and I is the direct current, then this expression is recognised simply as Ohm's law.

Also in a mechanical vibrating system when an external force F is applied and a velocity v is produced,

$$F/v = Z_m \qquad (1.9b)$$

called the **mechanical impedance**.

In the case of sound, when the sound pressure p is the action and produces a particle velocity v,

$$p/v = Z \qquad (1.9c)$$

is called the impedance for the sound wave at the point.

Strictly speaking, since in this case both the sound pressure and particle velocity are considered over unit area, Z in Equation (1.9c) is the **specific acoustic impedance** or **acoustic impedance density**. In a pipe whose cross-sectional area is S, a volume velocity Sv is produced by a sound pressure p,

$$p/Sv = Z_A \qquad (1.10)$$

called the **acoustic impedance** (see Section 11.2). These impedances relating to vibration are expressed by complex quantities because Action and Effect are generally not in the same phase, thus

$$Z = R + jX = |Z|\, e^{j\varphi} = |Z| < \varphi \qquad (1.11)$$

where, the absolute value of $|Z| = \sqrt{R^2 + X^2}$, and the phase angle $\varphi = \tan^{-1}(X/R)$.

When a plane wave propagates in free space through a medium without any loss, the acoustic impedance is always a real number that is a product of the medium density ρ and sound speed c (see Section 11.1),

$$Z = p/v = \rho c \qquad (1.12)$$

This Z has a specific value for the medium and is called the **characteristic impedance**, or **specific acoustic resistance** as it is a real number. For air in its standard condition the value is $415\ \mathrm{kg\ m^{-2}\ s^{-1}}$ (the unit $\mathrm{kg\ m^{-2}\ s^{-1}}$ is called a MKS rayl). Characteristic impedance values for various materials are shown in Table 1.1. The reciprocal of impedance is called **admittance**.

1.4   Sound intensity and level

A. *Sound intensity*

The energy passing through unit cross-sectional area normal to the direction of sound propagation in unit time (1 s) is termed the **sound intensity**. This is regarded as power per unit area. As mentioned above, when sound pressure is substituted for electric voltage and particle velocity for electric current, sound intensity corresponds to electric power, since the sound transmission path is analogous to the electric circuit. As is well known, electric power = electric voltage × electric current, therefore sound intensity I can be expressed as follows

$$I = pv \ (\mathrm{W\,m^{-2}})$$ (1.13)

From Equation (1.12)

$$I = p^2/\rho c = \rho c v^2$$ (1.14)

Thus, clearly sound intensity is proportional to the square of the sound pressure and the particle velocity, respectively.

Here, in Equations (1.13) and (1.14) both sound pressure and particle velocity should be effective values, as in the case of the electric alternating current. The **effective value** is the root mean square (RMS) of the instantaneous values as indicated in Figure 1.2, where there is no minus sign because of squaring. In the case of a sinusoidal vibration with amplitude A and period T, the RMS value is as follows:

$$\mathrm{RMS} = \sqrt{\frac{1}{T} \int_0^T (A \cos \omega t)^2 \mathrm{d}t} = \frac{A}{\sqrt{2}} = 0{\cdot}707A$$ (1.15)

Unless otherwise stated, the magnitude of the sound pressure and the particle velocity are shown as effective values.

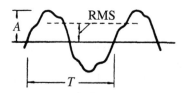

Figure 1.2 Effective values.

B. Sound energy density

The sound intensity of a plane wave I is equal to the energy travelling a distance c (m) through a unit area for 1 s, therefore, the sound energy density E in this space will be,

$$E = I/c = p^2/\rho c^2 \ (\text{W s m}^{-3} = \text{J m}^{-3})\qquad(1.16)$$

where c is the sound speed. Energy density is used since there is no need to consider direction, and determination of a sound field becomes much simpler in a room in which reflected sounds are travelling in all directions.

C. Decibel scale

In practice, when the sound intensity or sound pressure is measured, a logarithmic scale is used in which the unit is the decibel (dB). It derives from the auditory sensation of man, which can perceive a very wide range of sound intensities, i.e. the maximum to minimum energy ratio is more than 10^{13}: 1. Also the sensation is logarithmically proportional to the intensity of the stimulus (Weber–Fechner's law: see Section 1.8C).

Originally, the scale expressed the logarithm of the ratio of two powers, W_1 to W_0, called the **bel**, but as it is too coarse a unit one-tenth of it is used, the **decibel (dB)**.

$$\text{Number of decibels} = 10\log_{10}\left(\frac{W_1}{W_0}\right) = 10\log_{10}(n)(\text{dB})\qquad(1.17a)$$

It is used not only to make relative comparisons but also to express absolute values by reference to a standard value. With a standard value of $W_0 = 10^{-12}$ watt, the sound of W (watt) is expressed as

$$\text{Sound power level,}\quad L_W = 10\log_{10}\left(\frac{W}{10^{-12}}\right)(\text{dB})\qquad(1.17b)$$

A magnitude on such a logarithmic scale is generally called a **level**.
Thus, the **sound intensity level** of I (W m^{-2}) is expressed as follows,

$$\text{Sound intensity level} = 10\log_{10}\left(\frac{I}{I_0}\right)(\text{dB})\qquad(1.18)$$

where, $I_0 = 10^{-12}$ W m^{-2}.
Then from Equation (1.14) the **sound pressure level** L_p is obtained as follows:

$$L_p = 10\log_{10}\left(\frac{p^2}{p_0^2}\right) = 20\log_{10}\left(\frac{p}{p_0}\right)(\text{dB})\qquad(1.19)$$

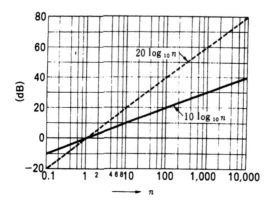

Figure 1.3 Decibel scale.

where, $p_0 = 2 \times 10^{-5}\,\mathrm{N\,m^{-2}} = 20\,\mu\mathrm{Pa}$ is the reference value for air. It is much easier to measure sound pressure than intensity, therefore, generally a sound field is expressed by the sound pressure level, which is, of course, equal to the sound intensity level for a plane wave in free space. In other instances it is common practice to evaluate the strength of sound in terms of the sound pressure level. For the calculation of the sound field in a room, the sound energy density is often used and compared directly as follows

$$\text{Energy density level} = 10\log_{10}\left(\frac{E}{E_0}\right)(\text{dB}) \qquad (1.20)$$

where an arbitrary value can be used for E_0 for purposes of convenience in calculation. As described above, although the value is expressed in dB, the reference value used should be clearly defined.

[Ex. 1.1] $L_1 = 10\log_{10} n(\text{dB})$ and $L_2 = 20\log_{10} n\,(\text{dB})$ are plotted graphically in Figure 1.3.

D. Energy summation and average using decibels

(a) Suppose we wish to find the level of sound L_3 (dB) resulting from two sounds L_1 (dB) and L_2 (dB), which exist simultaneously and comprise noise of random frequencies. Let their energy densities be E_1 and E_2, respectively. Then since $E_3 = E_1 + E_2$

$$L_3 = 10\log_{10}\left(\frac{E_3}{E_0}\right) = 10\log_{10}\left(\frac{E_1 + E_2}{E_0}\right)$$
$$= 10\log_{10}\left(10^{L_1/10} + 10^{L_2/10}\right)\text{dB} \qquad (1.21)$$

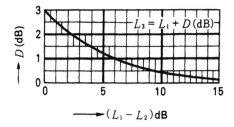

Figure 1.4 Summation of L_1 (dB) and L_2 (dB), $L_1 > L_2$.

This calculation is made easier by using Figure 1.4. However, when L_1 and L_2 (dB) have the same or nearly the same frequency content as pure tones, the above equation cannot be used because of interference (see Section 1.6); the reader must beware of the nature of the sound sources.

[Ex. 1.2] When the energy summation L_3 is obtained from the expression $L_3 = L_1 + D$ (dB), with above notation, the value of D vs $(L_1 - L_2)$ can be obtained from Equation (1.21) with the proviso that $L_1 > L_2$ as follows

$$D = 10\log_{10}(1 + 10^{-(L_1 - L_2)/10})$$

and Figure 1.4 is derived.

(b) When there are n sources of sound of L_n (dB) with energy density E_n, the average level $\overline{L}$ is given by

$$\overline{L} = 10\log_{10}\left(\frac{E_1 + E_2 + \cdots + E_n}{nE_0}\right) \text{dB} \qquad (1.22a)$$

or in terms of sound pressure p_n,

$$\overline{L} = 10\log_{10}\left(\frac{p_1^2 + p_2^2 + \cdots + p_n^2}{np_0^2}\right) \text{dB} \qquad (1.22b)$$

In practice

$$\overline{L} = 10\log_{10}\frac{1}{n}(10^{(L_1/10)} + 10^{(L_2/10)} + \cdots + 10^{(L_n/10)})\text{dB} \qquad (1.22c)$$

is used.

When the difference between the maximum and minimum values among the L_n is not greater than 3 (or 5) dB, the arithmetic mean can be used within an error of 0.3 (or 0.7) dB in place of $\overline{L}$.

1.5   Reflection, absorption and transmission

A. *Absorption coefficient and transmission loss*

When a sound hits a wall, its energy is divided into three parts. If the sound incident on a wall has energy E_i, a part of the sound energy E_r is reflected back while a part E_a is absorbed by the wall. The rest E_t is transmitted as shown in Figure 1.5. So we may write

$$E_i = E_r + E_a + E_t$$

Then, the **absorption coefficient** α is defined as follows:

$$\alpha = \frac{E_i - E_r}{E_i} = \frac{E_a + E_t}{E_i} \tag{1.23}$$

which means that all portions of the sound, except that which is reflected, is considered to be absorbed. The **transmission coefficient** τ is defined as follows

$$\tau = \frac{E_t}{E_i} \tag{1.24}$$

This can be expressed in dB and is called transmission loss or sound reduction index, denoted by TL or R, respectively, and defined as follows

$$\text{TL (or } R) = 10 \log_{10} \frac{1}{\tau} = 10 \log \frac{E_i}{E_t} \text{dB} \tag{1.25}$$

In practice this is the term commonly used to describe sound insulation.

[**Ex. 1.3**] For an open window, as Figure 1.5, then $E_r = 0$, $E_a = 0$. Therefore, $E_i = E_t$ and the absorption coefficient $\alpha = 1$. The transmission coefficient $\tau = 1$, which means complete transmission, therefore, TL $= 0$ dB. So in this case it is seen that a material that has a large absorption coefficient is no good for sound insulation.

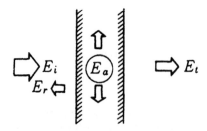

Figure 1.5 Reflection, absorption and transmission.

B. Reflection, refraction and diffraction

(a) A plane sound wave incident on a boundary between two different media shows the same phenomena of **reflection** and **refraction** as an optical wave. In Figure 1.6(a), the incident angle θ_i = reflection angle θ_r. With angle of refraction θ_t,

$$c_1 \sin \theta_1 = c_2 \sin \theta_i \tag{1.26}$$

If $c_1 < c_2$, and when

$$\theta_i \geqq \sin^{-1}(c_1/c_2) \tag{1.27}$$

then **total internal reflection** occurs. These relations are explained by **Huygens' principle**.

On the basis of these rules, the sound paths and wave fronts are drawn graphically in Figure 1.6(b) where I_1 and I_2 are the mirror images of the sound. The line showing the sound path is called a **sound ray** and the surface connecting vibrating particles in the same phase is called a **wave front**. The wave front created by the sound waves emerging from the sound source after t seconds forms a sphere of radius ct with its centre at the source of mirror image. However, for valid application of this reflection rule the reflection plane should be sufficiently large compared to the wavelength.

(b) Even if an obstacle exists in the space in which the sound wave propagates, the sound wave goes behind the body as shown in Figure 1.7. This phenomenon is called **diffraction**. The sound attenuation due to the diffraction depends on the size of the obstacle; the smaller the sound wavelength compared to the body the larger the attenuation and vice versa. The attenuation will be obtained by the wave theory described later (see Section 11.14). In the case of an uneven reflecting surface, although when the wavelength is

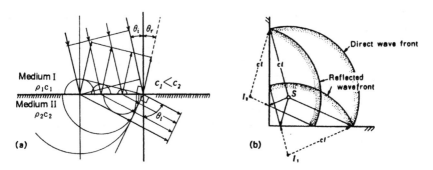

Figure 1.6 (a) Reflection and refraction drawn by Huygens' principle. (b) Wave fronts of direct and reflected sounds.

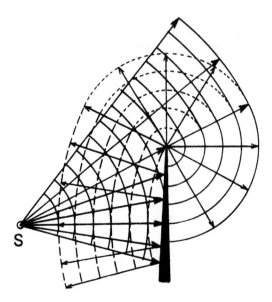

Figure 1.7 Sound reflection and diffraction by a screen.

very small the sound reflects from each portion of a concave or convex surface just as in optics, there is almost no effect when the size of a concave or convex portion is small compared to the sound wavelength. In optics, since the wavelength of light is very small, a distinct shadow is produced and reflected paths follow a graphical drawing, whereas in acoustics, because the wavelength of sound is relatively large (in the audible range from 2 cm to 17 m) and is close to the sizes of objects in our environment, the sound field is very complicated.

C. *Coefficients of reflection, transmission and absorption*

(a) Suppose the boundary plane between two media is infinitely large and a plane sound wave travelling in medium I is normally incident on the plane, as shown in Figure 1.8, where $\theta_i = \theta_r = 0$ (as in Figure 1.6(a)), then the relationships between the magnitude of the reflected and the transmitted sound is derived as follows.

At the boundary, two conditions of continuity must be fulfilled on both sides, i.e. (1) there must be continuity of sound pressure and (2) the particle velocities are equal. Though sound pressure has no direction, velocity is a vector quantity hence

$$\begin{cases} p_i + p_r = p_t \\ v_i - v_r = v_t \end{cases}$$

$$(1.28\text{a})$$
$$(1.28\text{b})$$

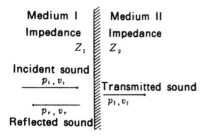

Medium I	Medium II
Impedance	Impedance
Z_1	Z_2

Figure 1.8 Reflection and transmission at the boundary plane between two media.

Both equations must be satisfied simultaneously. Substituting Equation (1.12) into Equation (1.28b)

$$\frac{p_i}{Z_1} - \frac{p_r}{Z_1} = \frac{p_t}{Z_2} \tag{1.29}$$

where $Z_1 = \rho_1 c_1$ and $Z_2 = \rho_2 c_2$.

Eliminating p_t from Equations (1.28a) and (1.29), the **sound pressure reflection coefficient** r_p is given by

$$r_p = \frac{p_r}{p_i} = \frac{Z_2 - Z_1}{Z_2 + Z_1} \tag{1.30}$$

Similarly, eliminating p_r from the equations, the **sound pressure transmission coefficient** t_p is given by

$$t_p = \frac{p_t}{p_i} = \frac{2Z_2}{Z_1 + Z_2} \tag{1.31}$$

Therefore, the two coefficients are determined from the characteristic impedances.

[Ex. 1.4] From the above conditions, the reflection coefficient r_v and transmission coefficient t_v can be obtained in a similar manner from the particle velocity as follows:

$$r_v = \frac{v_r}{v_i} = \frac{Z_2 - Z_1}{Z_2 + Z_1} \tag{1.32a}$$

$$t_v = \frac{v_t}{v_i} = \frac{2Z_1}{Z_1 + Z_2} \tag{1.32b}$$

(b) When considering the relationship between the aforementioned absorption coefficient and sound pressure reflection coefficient, since the reflected incident sound energies in Equation (1.23) are proportional to $|p_r|^2$ and $|p_i|^2$, respectively

$$\frac{E_r}{E_i} = \frac{|p_r|^2}{|p_i|^2} = |r_p|^2$$

Therefore,

$$\alpha = 1 - |r_p|^2 = 1 - \left|\frac{Z_2 - Z_1}{Z_2 + Z_1}\right|^2 \tag{1.33}$$

When medium I is air, $Z_1 = \rho c$, and medium II is an absorptive material, $Z_2 = Z$, the absorption coefficient of the material becomes,

$$\alpha = 1 - \left|\frac{(Z/\rho c) - 1}{(Z/\rho c) + 1}\right|^2 = 1 - \left|\frac{z - 1}{z + 1}\right|^2 \tag{1.34}$$

Thus, the absorption coefficient is determined by z, the ratio of the impedance Z to the characteristic impedance of the air.

$Z/\rho c = z$ is called the **acoustic impedance ratio** or **normalised impedance** of the material surface.

1.6   Interference, beats and standing waves

A. *Interference and beats*

When two waves of the same frequency propagate simultaneously, the vibration amplitude at each point of the medium is determined by the summation of the amplitudes of each individual wave. This is the principle of superposition in a linear system. Therefore, the amplitude increases where the waves meet in phase and decreases where they are out of phase. Such phenomena, where more than two sound waves overlap and cause this amplitude change, are called **interference**. When a pure tone is generated in a room, multiple reflected sounds are propagated from various directions, and as a consequence very complicated **interference patterns** are produced in the space.

A phenomenon known as **beats**, which is produced by two sounds whose frequencies are slightly different, is due to an amplitude pulsation, i.e. a change in time recognised at an observation point. The number of beats per second is equal to the difference in frequency of the two sounds. This phenomenon is used to measure precisely or tune the frequency of a sound when the other sound's frequency is recognised as a standard.

B. *Standing waves*

When two sound waves of the same frequency are travelling in opposite directions, the further from a reference point, then the more the phase shifts back and forth between the two waves. Therefore, at a certain point, where the waves are in phase, the amplitude becomes a maximum, producing an **antinode** while at another distance the amplitude becomes a minimum, producing a **node** with the waves out of phase. Thus, antinode and node are observed alternately so the identical motion repeats at every point and the wave form does not proceed. This phenomenon is called a **standing wave,** which is the simplest example of interference. A complicated interference pattern produced in a three dimensional room is also called a standing wave since the wave form does not move.

Expressing the sound pressure of a plane wave incident normally on a hard wall as

$$P_i = A \sin(\omega t + kx)$$

and for the reflected wave as

$$P_r = B \sin(\omega t - kx)$$

then, when they are superposed

$$P = P_i + P_r = A \sin(\omega t + kx) + B \sin(\omega t - kx)$$
$$= (A + B) \sin \omega t \cos kx + (A - B) \cos \omega t \sin kx$$

For simplification, assume that the sound is completely reflected, i.e. $A = B$,

$$P = 2A \sin \omega t \cos kx \qquad (1.35)$$

Since $\cos kx$ is the function expressing the form of the wave in space, when $kx = x(2\pi/\lambda) = n\pi$, i.e. $x = n(\lambda/2)$, $(n = 0, 1, 2, 3, ...)$ the sound

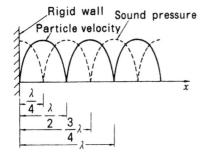

Figure 1.9 Standing wave.

pressure becomes a maximum and when $kx = x(2\pi/\lambda) = (2n+1)\,\pi/2$, i.e. $x = (2n+1)\,\lambda/4$ the sound pressure becomes zero as shown in Figure 1.9. This is the sound pressure pattern of a standing wave.

[**Ex. 1.5**] In order to obtain the particle velocity pattern of a standing wave, when a plane wave is incident on a hard wall, a similar expression to sound pressure can be used. The velocity pattern is as shown in Figure 1.9 by the solid curve, with zero velocity at the position of maximum sound pressure and maximum velocity at zero pressure.

1.7   Reverberation

When a sound source starts to supply sound energy in a room, it needs some time to build up an equilibrium sound level. Afterwards, if the sound source stops, the sound will still be heard for some time until it decays away completely as shown in Figure 1.10. Such a phenomenon, in which the sound remains even after the termination of the source, is called **reverberation**.

To evaluate reverberation numerically, the **reverberation time** is defined as the time required for the average sound energy density to decay by 60 dB from an equilibrium level. Since the time of W.C. Sabine in 1900 (Lit. B2), who studied the phenomenon, reverberation time has been used as the most important indicator of the acoustic characteristics or the auditory environment of a room. Sabine found, from many experimental results, that the larger the volume V (m³) of the room, the longer the reverberation time T, and the more absorptive the materials and objects in the room, the shorter the reverberation time. Therefore,

$$T = K\frac{V}{A}\text{(s)} \tag{1.36}$$

where the constant $K = 0.16$ and A is the total absorption. A can be obtained as follows

$$A = S\bar{\alpha}\text{(m}^2) \tag{1.37}$$

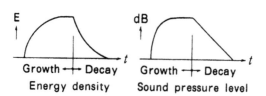

Figure 1.10 Sound growth and decay in a room.

where S (m^2) is the total surface area inside the room and the average absorption coefficient is $\bar{\alpha}$. The total absorption is expressed by

$$A = \Sigma S_i \alpha_i (\text{m}^2) \qquad (1.38\text{a})$$

where α_i is the absorption coefficient of the corresponding area S_i and $\Sigma S_i \alpha_i$ is the sum of the absorptions of all the different surfaces in the room.

When there are furniture and people present for which the surface area cannot be determined, each individual unit of absorption described as A_j (m^2) is used. Therefore, the total absorption can be obtained as follows:

$$A = \Sigma S_i \alpha_i + \Sigma A_j (\text{m}^2) \qquad (1.38\text{b})$$

This value A is called **equivalent absorption area**. α_i is variable depending on the frequencies of sounds, as shown in Table A.2 in the Appendices, however, generally the value at 500 Hz is taken as representative. There will be more detailed discussion in Chapters 3 and 4.

1.8   Loudness and loudness level of sound

Loud or silent describe a sound that produces a large or small auditory sensation, which is related to the physical intensity of the sound. However, the subjective sensation is not in simple proportion to the objective intensity, so another scale needs to be used.

A. Loudness level

The loudness level of a sound is defined as the sound pressure level in dB of a standard frequency, 1000 Hz, pure tone, which is heard with loudness equal to that of the sound, and measured in **phon**. The curves shown in Figure 1.11 that connect the equal loudness levels of each sound pressure level for each frequency of pure tone are called **equal loudness contours**.

The figures are based on the average values of tests carried out with numerous young persons aged from 18 to 20 years. We can observe the following in Figure 1.11.

1   Generally speaking, below 500 Hz the auditory sensitivity is reduced with decreasing frequency. At 100 Hz, for example, on the curve for 20 phon the sound pressure level is nearly 30 dB higher than at 1000 Hz, which means the sound energy is 1000 times that of the standard; at 40 Hz, 100 000 times; and at 20 Hz, about 10 000 000 times that of the 1000 Hz standard.
2   This fall in sensitivity becomes less pronounced at higher levels of the range.
3   Maximum sensitivity exists between about 3000 or 4000 Hz at the lower level, it reaches 3–6 dB from the standard.

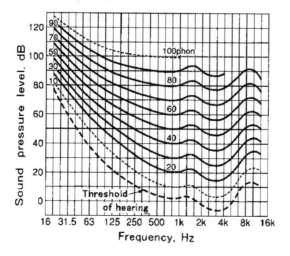

Figure 1.11 Normal equal-loudness-level contours for pure tones (binaural free-field listening and frontal incidence). Reproduced from ISO 226:2003 with the permission of the International Organization for Standardization, ISO. ISO standards and related documents can be obtained from the Japanese Standards Association (ISO member) at http://www.jsa.or.jp and from the ISO Central Secretariat at http://www.iso.org. Copyright remains with ISO.

B. Loudness

The loudness level can define sounds that are heard as equally loud, but cannot be used to compare directly sounds at different levels. For example, a sound of 100 phon is not heard as twice as loud as one of 50 phon. As a result of experiments carried out by Fletcher, a relationship was established between the loudness level in phons and the subjective loudness in **sones**. The unit of loudness is a sone, which is defined as the loudness of a 1000 Hz tone of loudness level 40 phon. If the sound is heard twice as loud as 1 sone, its loudness equals 2 sone, and if 10 times as loud the loudness is 10 sone. The relationship between loudness and the loudness level is obtained by experiments and standardised in Figure 1.12.

C. Weber–Fechner's law

As shown in Figure 1.11, the maximum to minimum energy ratio in the audible sound pressure level range is as much as 10^{13}:1. Due to Weber–Fechner's law, however, the **just noticeable difference (JND)** in the physical stimulus for the smallest change of human sensation, called the **differential limen**, is proportional to the original amount of physical stimulus, i.e. not

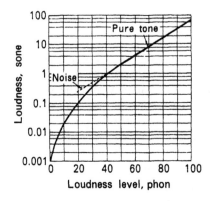

Figure 1.12 Loudness (sone) vs loudness level (phon).

only in hearing but all sensations (magnitude) are logarithmically proportional to the intensity of the stimulus. Therefore, the decibel scale can be applied to the measurement of sound and is also very convenient for the handling of large numbers.

1.9   Auditory sensation area

Ears can hear a wide range of sound pressures and frequencies as shown by the equal loudness contours in Figure 1.11. The minimum sound pressure that is perceived as a sound is called the **minimum audible field (MAF)** indicated by a broken line in the figure. The maximum audible value is not so clearly defined; a sound pressure level higher than 110 dB causes an uncomfortable sensation in the ear and at still higher levels, a feeling of pain. There is a threshold of feeling beyond which the sound is perceived not as sound but as pain, and its intensity level is about 130–140 dB for almost all frequencies and is accompanied by the risk of irreparable nerve damage.

The differential limen in sound intensity is about 1 dB, although it varies in a complicated way depending on frequency and sound intensity, and also on the hearing ability of the individual person.

The range of frequencies which can be heard as a sound is shown in Figure 1.11 and extends from about 20–20 000 Hz. The differential limen in frequency is about 0.7% above 500, and 3–4 Hz below 500 Hz, but may vary depending on the sound intensity and also on the individual person.

This auditory sensation area is based on tests with many young people who have good hearing. However, for people over 20 years old, the auditory sensation shows a noticeable falling off especially at high frequencies with increasing age. That is, the threshold of hearing, MAF, shifts upward in Figure 1.11. The amount of this shift is called **hearing loss** and is measured in dB.

1.10   Pitch and tone

A. *Sound scale (pitch)*

We have the ability to sense whether the pitch of a sound is high or low. This perception of the pitch of a sound depends mainly on the sound's frequency. It is, however, rather complicated since the intensity and the wave shape may affect our sensation. A sound that has a constant pitch is called a **tone**. When the duration is too short, we may not perceive the pitch.

Weber–Fechner's law may also be applied to frequency perception. The $\log_2(f_2/f_1)$ is defined as the octave number, i.e. one octave is where f_2 is twice f_1. This is the basis of the musical scale where we have a similar sensation when the frequency is doubled. In engineering we always identify the octave by its position on the frequency scale, though in psychoacoustics the pitch scale is in **mel**, a number that permits summation in a similar way to sone in loudness.

[Ex. 1.6] The frequencies of sounds at octave intervals, above and below 1000 Hz are, for example, 500, 250 and 125 Hz for the lower ones and 2000, 4000 and 8000 Hz for the higher. These frequencies are equally spaced on a logarithmic scale.

The sound generated from a musical instrument has many component tones of various frequencies and is called a **complex tone**. The tone of the lowest frequency is called the **fundamental** and all other tones are known as **overtones**. If the frequencies of the overtones are an integral multiple of the fundamental, they are called **harmonics**. The pitch of a complex sound is perceived as that of the frequency which is the highest common factor of the frequencies of all the component sounds. Therefore, a complex sound whose overtones consist of only harmonics has a pitch which is that of the fundamental even if that fundamental is actually absent.

[Ex. 1.7] The pitch of a complex sound consisting of components 100, 150, 200 and 300 Hz is, in fact, the pitch of 50 Hz which is the highest common factor for the components. Thus, even a loudspeaker that cannot produce a frequency less than 100 Hz, has a pitch equivalent to 50 Hz. This 50 Hz is called the **missing fundamental**.

B. *Timbre and spectrum*

We can distinguish the difference between tones produced by different musical instruments, even if they have the same pitch and intensity, because they possess their own **timbre**. A **pure tone**, which is represented by a pure sine wave, produces the simplest timbre while a complex tone changes the timbre depending on its component tones. The timbre can be physically compared by means of the **spectrum** of the sound obtained by frequency analysis.

The sound spectrum is measured by the **spectrum level**. The spectrum level of the frequency f is the sound intensity level in a 1 Hz bandwidth.

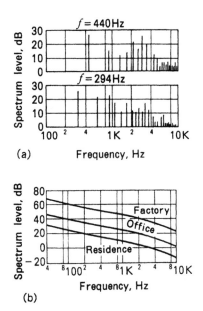

Figure 1.13 Examples of sound spectra: (a) spectrum of violin sound; and (b) spectrum of long-term average of environmental noise.

The spectrum of a musical tone is discontinuous as shown in Figure 1.13(a) because it consists of harmonics, while the spectrum of noise is generally continuous as shown in Figure 1.13(b).

The sound whose spectrum level is equal and uniformly continuous for the whole range of frequencies is called **white noise**, analogous to the optical phenomenon of white light, which has a continuous spectrum.

When a noise is to be analysed for noise control purposes, generally a frequency bandwidth of 1 or 1/3 octave is used. Thus the measured sound intensity level in the frequency band is called the 1 or 1/3 octave band level, and the result is described as a **band spectrum**.

Although we have no pitch sensation for white noise itself, we sense the pitch of the various bands filtered with 1 or 1/3 octave band filters. These band noises are often used as source signals for acoustic measurements.

[**Ex. 1.8**] In Figure 1.14 the centre frequency f_m for the octave and 1/3 octave band and the bandwidth Δf have the following relationships, for a 1 octave band

$$f_2 = 2f_1, \qquad \frac{f_m}{f_1} = \frac{f_2}{f_m} = \frac{2f_1}{f_m}$$

$$f_m = \sqrt{f_1 f_2} = \sqrt{2}f_1, \qquad \Delta f = f_1 = \frac{1}{\sqrt{2}}f_m = 0 \cdot 707 f_m$$

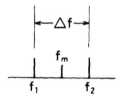

Figure 1.14 Frequency band: Δf, band width; f_m, centre frequency; f_1 and f_2, cut-off frequencies.

and for a 1/3 octave band

$$\frac{f_m}{f_1} = \frac{f_2}{f_m} = 2^{1/6}$$

$$\Delta f = f_2 - f_1 = f_m(2^{1/6} - 2^{-1/6}) = 0 \cdot 23 f_m$$

[**Ex. 1.9**] The relationships between spectrum level, octave band level and 1/3 octave band level for a white noise with intensity $I_s/1\,\mathrm{Hz}$ at frequency f are derived as follows for the spectrum level

$$L_s = 10 \log_{10} \frac{I_s}{I_0}\,(\mathrm{dB})$$

for the octave band level

$$L_1 = 10 \log \frac{\Delta f I_s}{I_0} = 10 \log \frac{I_s}{I_0} + 10 \log \Delta f = L_s + 10 \log 0 \cdot 707 f_m$$

for the 1/3 octave band level

$$L_{1/3} = L_s + 10 \log 0 \cdot 23 f_m$$

Thus, the band level is raised by 3 dB/octave when the frequency is increased in any relative bandwidth, since the intensity is proportional to f_m.

C. Timbre of a musical tone

Although the highest fundamental tone for the musical scale is about 4000 Hz, the timbre of each individual instrument is different depending upon the harmonic structure, which may include very high frequencies as shown in Figure 1.13(a). Furthermore, there are noise-like sounds from musical instruments due to frictional noises, such as the start of the passage of the bow across a violin string, the spectrum shift during the starting transient in the build up, and the decay of sound in a piano string and so on, all of which contribute to what we call timbre. The frequency range necessary to transmit exactly the timbre is generally from 30 to 16 000 Hz, i.e. almost the whole audible frequency range depending on the instrument.

Furthermore, to produce real high fidelity sound, it is believed that there is a need for sounds higher than what is audible, i.e. the ultrasonic frequency range.

1.11   Auditory masking

When we want to hear a sound in a noisy environment, we often experience difficulty hearing the sound clearly. So, we can say that the noise masks the sound, a phenomenon referred to as the **masking effect**. The explanation is that the noise dulls the auditory sensation, and as a result the threshold of hearing is raised. The amount of masking is measured by this threshold shift in dB.

A. Masking due to pure tones

Figure 1.15(a) is an example of an actual measurement and shows that a pure tone of 1200 Hz, with sound pressure level of 100 dB, the *masker*, raises the threshold of hearing another pure tone, the *maskee*, up to 70 or 80 dB

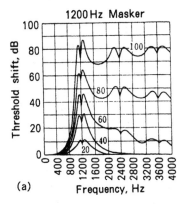

(a)

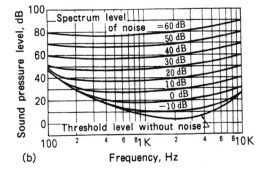

(b)

Figure 1.15 Auditory masking: (a) masking due to pure tones; and (b) masking due to white noise.

in the frequency range above 1000 Hz. Similarly for pure tones masked by other frequencies, so that:

1   The louder the masker, the stronger the masking effect. And in a frequency range higher than the masker, we perceive stronger masking than in a frequency range lower than the masker.
2   The closer the masker frequency, the stronger the masking of the maskee frequency. In closest proximity, however, we can easily recognise the maskee due to audible beats.

B. Masking due to white noise

The threshold of hearing is shifted due to the presence of white noise as shown in Figure 1.15(b) and the masking effect is almost uniform for all frequencies.

1.12   Binaural effects and sound localisation

We listen to music or a lecture with two ears. Unless a sound signal comes from a sound source in the median plane (which divides the human body into left and right halves), the input signals to the two ears are not identical, because of their location on the left and right sides of the head. Interaural amplitude and phase (i.e. arrival time) differences depend on the direction of arrival of the sound signal, because of the directionality of the head and pinnae. Thus, we can perceive the position, distance, spatial extent of the sound image (auditory event) and other spatial aspects of the sound, with our two ears. This is called the binaural effect (Lit. B25).

The most effective cues of directional localisation are interaural phase and amplitude difference in the frequency regions below and above 1500 Hz, respectively. In the horizontal plane the localisation blur (i.e. uncertainty of source localisation) is minimum (1–3°) for the frontal direction. For a source in a direction beyond 60° relative to the frontal direction, the localisation blur dramatically increases and it is as large as 40° for a source in the lateral direction (90° relative to the frontal direction). The localisation blur in the median plane is larger than that in the horizontal plane.

It is difficult to localise a pure tone, and especially it is impossible in a room because of a standing wave. However, it is possible to localise a complex tone, a click and a band noise, because they include overtones.

Sound localisation is sometimes possible even in a room where many reflections come from different directions. This phenomenon is called the **law of the first wave front** or the **precedence effect**. According to the law (or the effect), a sound image is localised in the direction of the sound wave that arrives first at the listener.

We can turn our attention to one speaker and understand what he is talking about even in noisy surroundings. This is called the **cocktail party** effect. But, if we occlude one ear, the conversation is much more difficult to understand.

Meanwhile, distance perception is not so well understood as directional hearing, though many studies of distance perception have been done (M. Morimoto).

1.13 Nature of speech

A. Spectrum and level distribution of speech

The spectrum of speech measured and averaged over a long-term interval at 1 m from the speaker's mouth is shown in Figure 1.16, and is the average for ordinary people. Since the figure shows the intensity per Hz, by integrating the whole range of frequency, the overall intensity level is obtained at about 65 dB. However, speech is a complicated sound whose intensity and spectrum are fluctuating with time, therefore, it is necessary to describe it statistically. Each syllable of speech lasts about 1/8 s so that the sampling interval of 1/8 s is said to be sufficient.

In Figure 1.17, the abscissa shows the relative values of the speech intensity level referred to the long-term interval average value, while the ordinate

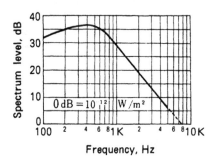

Figure 1.16 Long-term average spectrum of the human voice.

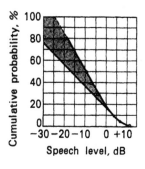

Figure 1.17 Ogive of speech level distribution at 1/8 s interval.

shows the time rate, i.e. the samples that exceed that level expressed as a percentage of the total. This characteristic is shown to be common to both male and females and also for any language. It shows that the long-term average level is only exceeded for 20% of the time while a level over +12 dB occurs about 1% of the whole time and the level difference between maximum and minimum levels (dynamic range) is found to be 45–55 dB. Referred to the average value, there is a variation from +40 dB for shouting to −40 dB as a minimum level.

B. Directivity of human voice

The sound intensity of the human voice has **directivity** caused by diffraction by the head as shown in Figure 1.18. Although over the whole frequency range, the difference is only −5 dB for the back of the head compared to the front. In the more important range for speech above 2000 Hz the difference is more than −15 dB.

C. Vowels and consonants

Human speech generally consists of **vowels** and **consonants**. The vowel sounds are produced by vocal cords during expiration, while the consonants are produced by the noise of air movement through the **vocal tract**, i.e. the throat, nostril and mouth over the tongue and lips. The frequency spectra of both types of sounds are formed by the resonant cavities in the vocal

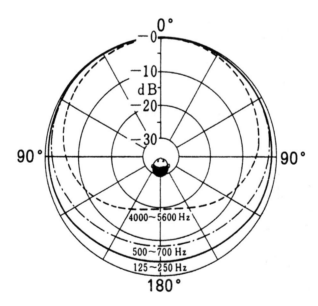

Figure 1.18 Directivity of human voice.

tract. The spectra of vowels have their own, almost constant, frequency ranges that are called **formant** frequencies. The pitch of the voice is based on the frequency of the vocal cords, i.e. about 100 Hz for male voices and 200–300 Hz for female voice as fundamental frequencies. It is an important characteristic that formants are almost stable, when the fundamental tone is varied.

The consonants are produced during the transient state of speech, so that the lapse time is very short and frequency components tend to be in the higher frequency range. The energy content in consonants is, unfortunately, so small that they are easily masked by noise, although they are critical for speech intelligibility.

1.14 Intelligibility of speech

The determination of speech intelligibility is important for evaluating an acoustic environment directly and synthetically, a technique that has its origins in the evaluation of telecommunication systems. Although many methods for evaluating room acoustics have been developed, as described in Chapter 3, in order to understand fundamental factors a classical measure is described here.

A. Percentage articulation

The percentage of meaningless spoken syllables correctly written down by listeners is called **percentage articulation (PA)**. Knudsen (Lit. B4) reported that on the basis of experiment, PA can be expressed as

$$PA = 96\, k_l k_r k_n k_s \ (\%) \tag{1.39}$$

where k_l, k_r, k_n and k_s are the coefficients determined by the average speech level, reverberation time, noise level, and room shape, respectively (see Figure 1.19).

k_l becomes a maximum at 60–80 dB and falls rapidly below 40 dB. k_r decreases when the reverberation increases. This is explained by the fact that the preceding sound overlaps and masks the succeeding one. Also the larger the noise level, the more the masking, so k_n decreases. In the figure the solid line shows the case where the speaker is situated in a noisy environment and deliberately controls his voice intensity, whereas the dotted line shows the situation where the speaker is in a different room, or the speech is recorded, so that there is no control of speech level. Although k_s is said to be defined by room shape, it may include other uncertain factors such as echo, etc. The fact that even when every coefficient is 1 indicating optimal conditions, PA is 96% instead of 100%, which is interpreted as the syllable itself possessing some obscurity.

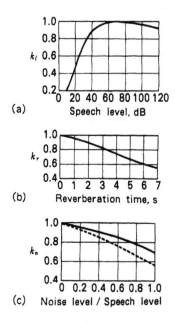

(a)

(b)

(c)

Figure 1.19 Reduction factors on percentage articulation: (a) speech level; (b) reverberation time; and (c) noise level.

B. Speech intelligibility

Even if the syllables are not fully recognised, the words or phrases are rather easily understood due to sequence or context. The percentage of correctly received phrases is called speech intelligibility. The relationship between syllable articulation and speech intelligibility is shown in Figure 1.20.

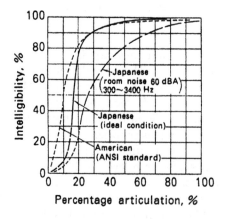

Figure 1.20 Speech intelligibility vs percentage articulation.

PROBLEMS 1

1   In a plane wave when the displacement of a medium is expressed as

$$\xi = Ae^{j\omega(t-x/c)},$$

show that the particle velocity $v = j\omega\xi$, and the acceleration $dv/dt = -\omega^2\xi$.

2   When a plane wave is incident on the boundary between two media at an angle of incidence, θ_i, show that the sound pressure reflection coefficient r_p can be expressed as follows

$$r_p = \frac{Z_2\cos\theta_i - Z_1\cos\theta_t}{Z_2\cos\theta_i + Z_1\cos\theta_t}$$

(Refer to Figures 1.6(a) and 1.8.)

3   There are a number of fans of the same type and size in a room. When a fan runs, the average sound pressure level in the room is 55 dB. When 2, 3, 4, ... and 10 fans run successively, calculate the average sound pressure levels at every stage in the room.

4   When a noise survey is carried out in a noisy factory, sound pressure levels of 58, 63, 59, 62, 64 and 61 dB are obtained at six points. Calculate the average sound pressure level in the room, and compare it to the arithmetic mean.

5   White noise has a spectrum level of 30 dB. Calculate octave band levels at the centre frequencies 100, 200, 500 and 1000 Hz. And also calculate 1/3 octave band levels for the same frequencies.

6   Describe fully the sound measuring units dB, phon and sone.

2 Noise and vibration measurement and rating

Any sound that a listener finds undesirable is defined as **noise**. Even beautiful music when it disturbs someone's study or sleep is perceived as noise. Vibration should also be considered as an environmental factor that brings disturbance to human comfort and activities.

2.1 Measurement of sound and noise

A. Sound level meter

The instrumentation must comply with specifications given in IEC 61672-1, which has two performance categories, classes 1 and 2. Both classes have the same design goals and differ only in the tolerance limits, class 1 for precision sound level meters, and class 2 for ordinary sound level meters. Although many countries have their own national standards, in general they follow the relevant International standards, those of the IEC: International Electrotechnical Commission.

a. Construction of sound level meter

A block diagram of a sound level meter is shown in Figure 2.1. A sound pressure omnidirectional microphone converts the sound pressure into an electrical voltage, which is amplified, passes through a frequency-weighting and a time-weighting network, which approximates to the ear's characteristics, and displays the results of the measurement. The display devices permit measurements with a resolution of 0.1 dB, or better, over the display range of 60 dB.

A sound level meter may include extensive signal processing, computers, recorders, printers and other devices.

b. Frequency weighting

Table 2.1 and Figure 2.2 show the frequency weightings A, C and Z, and corresponding tolerance limits for Classes 1 and 2. A and C weightings

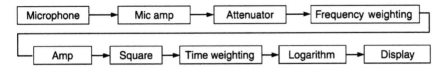

Figure 2.1 Block diagram of a sound level meter.

Table 2.1 Frequency weighting for sound level meter (from IEC 61672-1)

Frequency (Hz)	Frequency weightings (dB)			Tolerances (dB)	
	A	C	Z	Class 1	Class 2
16	−56.7	−8.5	0.0	+2.5; −4.5	+5.5; −∞
31.5	−39.4	−3.0	0.0	±2.0	±3.5
63	−26.2	−0.8	0.0	±1.5	±2.5
125	−16.1	−0.2	0.0	±1.5	±2.0
250	−8.6	0.0	0.0	±1.4	±1.9
500	−3.2	0.0	0.0	±1.4	±1.9
1000	0	0	0	±1.1	±1.4
2000	+1.2	−0.2	0.0	±1.6	±2.6
4000	+1.0	−0.8	0.0	±1.6	±3.6
8000	−1.1	−3.0	0.0	+2.1; −3.1	±5.6
16 000	−6.6	−8.5	0.0	+3.5; −17	+6.0; −∞

approximate the frequency response of the ear, 60 and 100 phon curves, respectively, as shown in the equal loudness contours (Figure 1.11). The intermediate weighting B, and weighting D proposed for rating aircraft noise in Figure 2.2, are no longer used.

The measured value dBA*, using the A-weighting, is regarded as a close approximation to the noise level perceived by human ears. The value of dBC*, using the C-weighting, may be taken as an approximate value of the

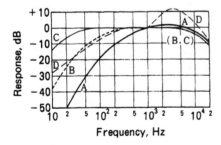

Figure 2.2 Frequency weighting characteristics.

*dB(A) or *dB(C) are also used, but omission of the parentheses is preferred throughout this book.

Table 2.2 Reference 4 kHz toneburst responses and tolerance limits for time-weighting

Time weighting	Toneburst duration, T_b (ms)	Max. response relative to the steady sound level (dB)	Tolerance limits (dB)	
			Class 1	Class 2
F (fast)	1000	0.0	±0.8	±1.3
	50	−4.8	±1.3	+1.3; −1.8
	2	−18.0	+1.3; −1.8	+1.3; −2.8
	0.25	−27.0	+1.3; −3.3	+1.8; −5.3
S (slow)	1000	−2.0	±0.8	±1.3
	50	−13.1	±1.3	+1.3; −1.8
	2	−27.0	+1.3; −3.3	+1.8; −5.3

sound pressure level measured by the Z-weighting. In noise surveys both dBA and dBC values should be recorded; then, from the difference between them, one can find which frequency range is dominant, either above or below 1 kHz without using a frequency analyzer.

c. Time weighting

The measured sound level is displayed by the RMS value of the signal, with F (fast) and S (slow) time-weightings. The averaging circuit has two time-constants (see Section 11.2.E): 125 ms for F and 1000 ms for S. The time weighting should also respond to 4 kHz tonebursts as specified in Table 2.2 (abbreviated from IEC 61672-1).

d. Correction for background noise

The reading on a sound level meter shows a total sound pressure level from many noise sources surrounding the microphone and is referred to as **ambient noise**. When only a **specific noise** emitted from a specific source is to be measured, the effect of residual noise must be excluded. The ambient noise remaining at the microphone position when the specific noise is suppressed is called **background noise** or **residual noise**. Unless the background noise level is sufficiently low compared with the one to be measured, proper measurement cannot be carried out. So, specifying the sound to be measured as S (signal), where $S = L_1$ (dB), the background noise N, where $N = L_2$ (dB), then $L_1 - L_2$ (dB) is called the *SN* **ratio** (signal-to-noise ratio). When S coexists with N, the measurable level L_3 (dB) may be obtained from Equation (1.21). Referring to Figure 1.4, when the *SN* ratio is greater than 10 dB, the background noise can be ignored.

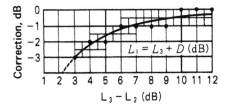

Figure 2.3 Correction to background noise for a specific noise measurement.

When L_2 and L_3 are measured, L_1 may be obtained from Figure 2.3 by the same process as in Figure 1.4. However, if (L_3-L_2) is smaller than 3 dB, the L_1 value may not be reliable.

[**Ex. 2.1**] In a factory the noise level including the noise generated by a machine is 92 dB and with the machine stopped it is 87.5 dB, so to obtain the machine noise level

$$L_3 - L_2 = 92 - 87.5 = 4.5 \, \text{dB}$$

From Figure 2.3 the correction is -2.0 dB. Hence,

$$L_1 = 92 - 2 = 90 \, \text{dB}$$

e. Other undesirable effects and their remedies

1   *Effect due to reflection and diffraction*: The measurement position should be at least 1 m from any wall, the ground and other objects. When an obstacle approaches the sight-line from source to microphone, a deviation may occur in the measured values. Moreover, the microphone shall be as far as possible from the observer, and if possible on a tripod or fixed stand.

2   *Wind effect*: When outdoors and at positions receiving wind, for example, when located near a blower system, it is difficult to get a correct value of the noise level due to wind-induced noise. A wind screen or windshield is essential in order to obtain correct measurements.

3   *Vibration effect*: Where subject to vibration, as in a vehicle or on board ship, it is advisable and effective to hold the microphone and the meter in the hand or support them on a flexible piece of plastic foam to reduce the vibration.

4   *Effect due to electromagnetic fields*: Measurements should not be carried out near an electric motor or transformer unless a condenser microphone is used. Caution is required in order to avoid induced currents in the microphone cable and other parts of the instrumentation.

5   *Effects due to temperature and humidity*: The standard conditions are a temperature of 20°C and 65% relative humidity. When the sensitivity changes by more than 0.5 dB with any other changes in condition, corrections must be provided by the manufacturer.

B. Frequency analysis

Frequency analysis of noise is indispensable for the purpose of not only realising the various effects of noise but also preventing them. Frequency analysis is performed by measuring the output of a sound level meter through a band filter that passes only a particular frequency range between f_1 and f_2 (Hz). In the analysing process $f_2 - f_1 = \Delta f$ (Hz) is called the **bandwidth** or **passband** where f_1 and f_2 are the **cut-off frequencies** and $f_m = \sqrt{f_1 \times f_2}$ is called the **centre** or **mid frequency**. There are two types of analysers, i.e. where the ratio $\Delta f / f_m$ is constant and where the bandwidth Δf is constant. Although the frequency component details become clearer as the filter bandwidth narrows, more time and effort are required for analysis; therefore, it is necessary to select the analyser which is most appropriate for the analysis of the particular noise.

a. Octave-band analyser

When the bandwidth is one octave, i.e. $f_2 = 2f_1$, the band filter is called an **octave-band filter**, while an **octave-band analyser**, which consists of several series of octave-band filters, is used for band-level measurements by switching to each serial band (see Table 2.3). Figure 2.4 shows an example of the standard octave-band filter performance specified by the IEC. The portable version of this analyser attached to a sound level meter is very simple and convenient to use for analysing noise in the field and on building sites.

b. 1/3 Octave-band analyser

Where more detailed analysis is required, 1/3 octave-band filters, which are produced by dividing the octave bandwidth into three equal parts (see Table 2.4), are used. The data obtained by a 1/3 octave-band analyser are

Table 2.3 Frequency of one octave band filter (Hz)

Centre frequency		31.5		63		125		250		500		1000		2000		4000		8000		16000
Cut off frequency	22.1		44.2		88.4		177		354		707		1410		2830		5660		11300		22600

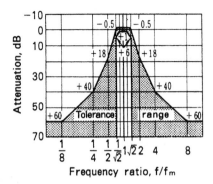

Figure 2.4 Frequency characteristics of octave-band pass filter (IEC 61260).

Table 2.4 Serial number used in 1/3 octave band centre frequency

1	1.25	1.6	2	2.5	3.15	4	5	6.3	8	10

often used in various evaluations required in architectural acoustics, mainly in laboratory measurements.

c. Real-time analyser

When the signals change rapidly in both amplitude and frequency, they can be processed simultaneously through all the filters covering the frequency range concerned by means of a real-time analyser and the output fed to a continuous display on a screen. The latter can display a series of instantaneous octave or 1/3 octave-band spectra, which are renewed many times per second. The output can then be recorded using any type of recorder or other type of printer. There are several types of compact sound level meters built for this kind of analyser on the market.

d. Narrow-band analyser and others

In some special cases in investigations of the noise source, a narrow bandwidth is required to obtain greater resolution. For this purpose, a fully digitised instrument based on FFT (Fast Fourier Transform, see Section 11.3) techniques is available. The time history of the sound is captured with an A–D (analogue–digital) converter and displayed directly on the screen. Quick transformations between the time and frequency domains are effected with a push-button, and the spectrum is displayed with both values of frequency and level in digital form, together with a graphical display on

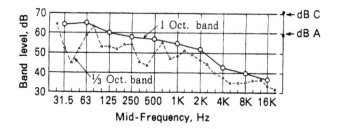

Figure 2.5 Frequency analysis for 1 and 1/3 octave bands.

the screen. This function can also be performed by a personal computer with the appropriate software.

e. Arrangements and conversion of analysed data

1 The analysed values should be plotted at the centre frequency of each particular band as shown in Figure 2.5. The bandwidth should be given for these plots.

2 The relationship between band level L_b and spectrum level L_s is expressed as follows:

$$L_s = L_b - 10 \log_{10} \Delta f \text{(dB)} \tag{2.1}$$

derived from [Ex. 1.9] based on the assumption that levels within Δf are uniformly continuous.

3 The total overall level L_{OA} is obtained from the various band levels $L_1, L_2, \ldots, L_n$, as follows

$$L_{OA} = 10 \log_{10} \frac{\sum P_n^2}{P_0^2} = 10 \log_{10} \left(\sum 10^{L_n/10} \right) \tag{2.2}$$

where P_n indicates the sound pressure in each band level. Hence, using Equation (1.21) and Figure 1.4, a successive calculation is carried out as follows: first, calculate the energy sum L_{12} for L_1 and L_2, then the sum for L_{12} and L_3 and so on.

4 In order to obtain dBA values, A-weighting values shown in Table 2.1 are added to all measured band levels, L_n, to obtain the **overall** total level.

C. Recording the data

In order to achieve accurate measurement and analysis, recording of the sound and vibration phenomena is a very important technique.

a. Level recorder

A level recorder captures the sound and/or vibration levels using a pen on a paper chart running at constant speed; the movement of the pen coincides with fluctuations in sound and/or vibration levels.

Historically, several types of level recorders were used (see Figure 3.16.), but they have disappeared from the market. This function has been succeeded by the digital processing of personal computers, as described in the next section, which use signal processing techniques, with results displayed on a screen and/or printed on paper.

b. Data recorder

In order to analyse a transient or irregularly fluctuating signal of sound and vibration, the output signal from the sound/vibration level meter is recorded on a magnetic tape-recorder or in the digital memory of a computer, and the recorded signal is played back whenever it is needed for analysis.

As there are many kinds of data recorder on the market, it is possible to choose a suitable one for any specified purpose. Also a stereo or multi-channel recorder for music recording can be used for sound analysis, but it is essential to calibrate the frequency characteristics and dynamic range of linearity for the measuring range of interest.

Digital memory devices have been widely developed, using the large capacity of solid-state memory, and data recorders can be constructed and executed using only electronic devices without any mechanical components. These are not only easier and more convenient for operating and maintaining, but also for installing other functions such as calculating any parameters with signal processing techniques.

D. Sound intensity measurement

With the rapid development of digital techniques, sound intensity measurements are now easily made with the aid of specially designed instrumentation systems. Sound intensity is a product of sound pressure and particle velocity as shown in Equation (1.13), which is a vector quantity as the particle velocity has both magnitude and direction. There are two principles upon which the design of a sound intensity probe is based; one uses a pressure-sensitive microphone and a velocity sensor, the other has two pressure microphones that are located at the proper distance for measuring the pressure gradient rather than particle velocity.

Both systems are fully digitised and the intensity measurement is carried out automatically. It is, however, necessary to take measurements at more points than are required for sound pressure measurements; also the directional characteristics have to be taken into consideration. Nevertheless, intensity measurements are now used for all kinds of acoustic assessments because the sound energy flow in any sound field can be visualised and so many acoustic quantities are defined or specified using sound energy as their basis: for example, the sound power of noise sources, absorption coefficient and transmission loss, etc. (see Lit. B38).

2.2   Noise rating (Lit. B24)

A. *Noise pollution*

Any sound undesired by the recipient is classed as noise, as it detracts from the quality of human life. An outline of the impact of noise on comfort is discussed next.

a. *Annoyance to daily life*

Because of masking due to noise, the comprehension of speech and music is affected due to a decrease in intelligibility (Figure 1.19). Background noise occurring in telephone, radio and television communication causes poor information transfer, affecting daily social life and productivity. Too much noise is psychologically unacceptable and causes annoyance and interferes with concentration, resulting in a decrease in efficiency leading to mistakes and errors. It also disturbs rest and sleep.

b. *Physiological effects*

Noise may cause temporary or permanent disorder in all the physiological functions of the digestive organs, respiratory organs, circulation and nervous systems. Ando & Hattori (1970) reported some effects on foetal life due to aircraft noise near an airport. It is becoming clear that such environmental noise affects the growth of children.

The most distinct physiological effect is hearing loss. When momentary hearing loss occurs due to exposure to short-duration, high-level noise, a rise in the threshold of audibility called **TTS** (temporary threshold shift) is observed, which may be recovered with time. However, people working for long hours in extremely noisy factories may risk permanent damage to their hearing referred to as **PTS** (permanent threshold shift).

c. *Social effects*

Alongside or near noisy trunk roads or busy airports, the demand for land use is reduced, resulting in lowering land values. Transport noise also affects not only human health but cattle and poultry growth, resulting in lawsuits for compensation because of decreases in milk or egg production. Thus, noise nuisance is becoming omnipresent in our society.

B. Steady noise rating

The effect of noise varies, largely depending not only on its intensity but on frequency and the time-varying pattern. As far as the frequency is concerned, noise whose major components are at high frequencies and/or containing pure tones appears noisier. As regards the time domain, generally an interrupted and/or impulsive noise is more annoying than a steady one. Even though the physical stimulus is identical, the effect will vary greatly depending on the listener's physiological and psychological state.

An ideal noise-rating system would be one in which all influencing factors are combined into a unique rating scale suitable for all kinds of noise. However, it is extremely difficult to unify the physical and physiological reactions to noise. Therefore, at present, depending on the ultimate aim, the following assessment methods are employed.

a. Loudness level (phon) and A-weighted sound pressure level (dBA)

When measuring the physical magnitude of the frequency spectrum of a steady sound, a couple of subjective judging methods are used, such as loudness **sones** and loudness level **phons** (see Sections 11.5, 11.6). Although they require much processing for evaluation, it has been shown that phon values correspond to L_A (dBA) measured by a simple sound level meter over a wide range of levels. So, the A-weighted sound level is now commonly used.

Table 2.5 shows acceptable value of dBA for various occupied environments.

b. Noisiness and PNL (PNdB)

Kryter (1970, Lit. B21) proposed a new subjective rating scale (mainly for the assessment of aircraft noise), which classifies **noisiness** in units of **noy**, following Steven's method relating loudness to units of sone (see Section 11.6). The noise level derived from this new scale is called PNL (perceived noise level) whose unit is PNdB (see Section 11.7). Since this method involves much processing, a simplified method using dBD values measured with a sound level meter with the frequency weighting D (Figure 2.2) had been proposed (1976), though it is not used now. For example, in the case of noise from jet aircraft

$$PNL = dBD + 7. \tag{2.3}$$

Alternately, using a dBA value measured with the usual A-weighting

$$PNL = dBA + 13 \tag{2.4}$$

is recognised as a better approximation to the perceived noise level.

Table 2.5 Recommended range of NCB curves (Figure 2.6) and dBA values for various occupied activity areas (Beranek 1988)

Type of space	NCB curve	dBA
Broadcast and recording studios (distant microphone used)	10	18
Concert halls, opera houses and recital halls	10–15	18–23
Large theatres, churches and auditoriums	<20	<28
Television and recording studios (close microphone used)	<25	<33
Small theatres, auditoriums, churches, music rehearsal rooms, large meeting and conference rooms	<30	<38
Bedrooms, hospitals, hotels, residences, apartments, etc. Classrooms, libraries, small offices and conference rooms	25–40	33–48
Living rooms and drawing rooms in dwellings	30–40	38–48
Large offices, receptions, retail shops and stores, cafeterias, restaurants, etc.	35–45	43–53
Lobbies, laboratories, drafting rooms and general offices	40–50	48–58
Kitchens, laundries, computer and maintenance shops	45–55	53–63
Shops, garages, etc. (for just acceptable telephone use)	50–60	58–68
For work spaces where speech is not required	55–70	63–78

But for other noise sources further research is still needed, since they depend mainly on frequency characteristics.

c. Assessment of frequency characteristics

1   *Speech Interference Level (SIL)*. In order to assess the disturbance caused in speech communication by noise, the arithmetic mean value of 4 octave-band levels, at mid-frequencies of 500, 1000, 2000 and 4000 Hz, is often used in the USA.

2   *NC and NCB curves*. Beranek (1957) developed Noise Criteria (NC) to deal with commercial buildings. The criteria consisted of a family of curves that related the noise spectrum to the disturbance caused in speech communication. These curves were later revised in 1988 as Balanced Noise Criteria (NCB) with improvements both at low and high frequencies, as shown in Figure 2.6.

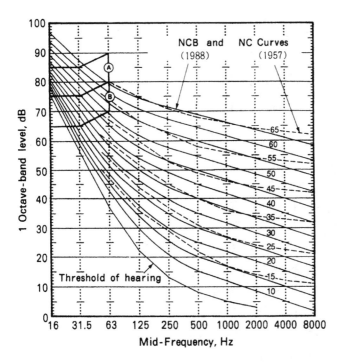

Figure 2.6 NC and NCB curves (Beranek 1988, Lit. B9).

The NCB number of a noise can be determined by plotting the noise in octave-band levels on the NCB curves. The lowest curve, which is not exceeded in any of the octave bands, is the NCB rating of the particular noise. The recommended values of NCB are given in Table 2.5 for the environmental assessment. This method is useful because any information relating to frequency, necessary for noise reduction purposes, can be easily extracted.

These values are applicable for background noise under the condition of all the facilities in operation in the room. Therefore, the permissible value of each noise source is obtained using the correction described in Section 2.1A.d.

3   *NR Curves.* The noise rating method was developed from the NC curves for wider application by a committee of the ISO (International Organization for Standardization) as shown in Figure 2.7. Though it was not accepted as an international standard, it is widely used in Europe for steady noise rating.

4   ***Hearing Damage Risk Criteria.*** The Japanese Industrial Hygiene Association (1969–2004) recommended the damage risk contours as shown in Figure 2.8, for simple continuous noise exposure instead of the criteria issued by the Committee on Hearing and Bioacoustics (CHABA)

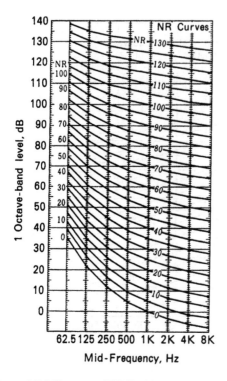

Figure 2.7 NR curves (ISO/R1996).

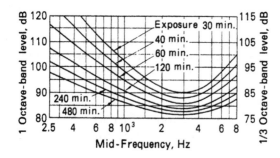

Figure 2.8 Hearing damage risk criteria (Japanese Industrial Hygiene Assoc., 1969–2004).

of the US Academy of Sciences. These contours indicate permissible exposure-minutes against band spectrum levels, which correspond to about 85 dBA in an 8-hour working day, so that the PTS will not exceed 10 dB at 1 kHz and below, 15 dB at 2 kHz and 20 dB at 3 kHz and above, for the average normal listener.

However, the assessment method for risk of noise-induced hearing impairment is being discussed continuously in the Committee of the ISO, since the more important recent concept is one of maximum permissible noise dose, which takes into account both the time-varying noise level and its duration.

C. *Time-varying noise measurements and rating*

Noise generally fluctuates with time, and its effect is highly dependent on its time-varying pattern. An intermittent or impact sound is judged as more annoying than a continuous sound. When the meter reading, however, fluctuates within a range of less than 5 dB when using the time-weighting S, then the noise can be treated as a steady noise and the average meter deflection taken.

a. *Sampling or statistical measurements*

1   *Discrete noise event having a relatively constant peak.* Noise, like that produced by a train, requires an average value from several measurements, each one taken over a specified period and time interval.
2   *Separate noise event having a wide range of peak values.* Noise, similar to that produced by aircraft, needs to be expressed in terms of the number of peak values that occur during the measuring period; for example, 13 times above 80 dBA, 10 times above 90 dBA, 3 times above 100 dBA, and the maximum value, 105 dBA, in an interval of 1 hour.
3   *Irregularly fluctuating noise with wide range in amplitude.* Noise, such as occurs in the street, can be obtained in terms of a **percentile level** $L_{AN,T}$, which is obtained by using the time-weighting F exceeded for N per cent of the time interval, T, as shown in Figure 2.9. For example, after reading 50 samples of instantaneous levels in the interval $\Delta t = 5$ s, sorting them into each level as shown in Figure 2.10, where the median value L_{50} lies at the centre of the ogive, then the 90% variation range, inclusive of 90% of the samples, is indicated by L_{95} and L_5 on the upper abscissa.

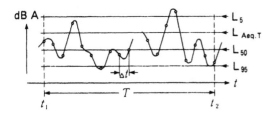

Figure 2.9 Irregularly fluctuating noise with wide dynamic range.

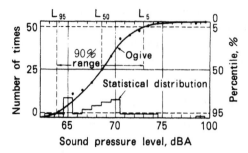

Figure 2.10 Ogive and percentile level of widely fluctuating noise.

b. Measurements based on sound dosage

Generally the effect of noise on human life can be taken as approximately proportional to the total energy of the existing noise stimulus. On the other hand, environmental noise is often a combination of sounds from many sources, and the distribution of such different kinds of sound is likely to change from moment to moment. Although there is a **dosimeter** measuring **total noise exposure level** on the market, the International Standard (ISO 1996-1) has defined the basic quantities, $L_{Aeq,T}$ and L_{AE} etc., in order to describe the noise in a community environment. The definitions are reviewed in the following:

1   *Equivalent continuous A-weighted sound pressure level.* This is the level in dBA of a continuous steady sound, which has the same A-weighted sound energy as the actual noise history within a specified time interval T. It is defined as

$$L_{Aeq,T} = 10\log_{10}\left[\frac{1}{t_2 - t_1}\int_{t_1}^{t_2}\frac{P_A^2(t)}{P_0^2}dt\right](dB) \qquad (2.5)$$

where $T = t_2 - t_1$, $P_A(t)$ is the instantaneous A-weighted sound pressure and P_0 is the reference sound pressure ($20\,\mu\mathrm{Pa}$).

The simplest method of measuring $L_{Aeq,T}$ is to use an integrating sound level meter (IEC 61672-1), which calculates and indicates the value of $L_{Aeq,T}$ together with the value of T automatically. Use can be made of an ordinary sound level meter by employing a sampling process. For this purpose $L_{Aeq,T}$ can be written as

$$L_{Aeq,T} = 10\log_{10}\left[\frac{1}{n}(10^{L_{A1}/10} + 10^{L_{A2}/10} + \cdots + 10^{L_{An}/10})\right](dB) \qquad (2.6)$$

where n is the total number of samples and L_{A1}, L_{A2}, ..., L_{An} are the measured sound levels in dBA.

When the sampling period Δt is shorter than the time constant of the measuring system, almost the same result is obtained as that obtained with the integrating sound level meter. However, in practice the recommendation is as follows

with time-weighting F $\Delta t \leq 0.25$ s

with time-weighting S $\Delta t \leq 2.0$ s

When the noise fluctuation is small, Δt can be longer by 5.0 s, then the conventional instrument is still useable.

If the noise level fluctuations form a normal distribution then the relation with the percentile level is as follows:

$$
\begin{aligned}
L_{Aeq} &= L_{A10} - 1.3\sigma + 0.12\sigma^2 \\
&= L_{A50} + 0.12\sigma^2
\end{aligned}
\tag{2.7}
$$

where σ is the standard deviation.

It is also said that L_{Aeq}, for highway traffic noise, is equivalent to $L_{A25} \sim L_{A30}$.

2 *Single event of a discrete noise.* The **sound exposure level** of a discrete noise event is defined as

$$
L_{AE} = 10\log_{10}\left[\frac{1}{t_0}\int_{t_1}^{t_2}\frac{P_A^2(t)}{P_0^2}dt\right](dB)
\tag{2.8}
$$

where t_0 is the reference duration of 1 s; $t_2 - t_1$ is the stated time interval, which is long enough to encompass all significant sound of a stated event.

An integrating sound level meter is a convenient instrument to use for this measurement. For an impulsive noise using an ordinary sound level meter, an approximate value of L_{AE} can be obtained from the peak value by using the S time-weighting.

When a noise event has a rather long duration of more than several seconds, with a sampling period Δt short enough to trace the noise-varying pattern, L_{AE} can be obtained using the conventional instrument with the S time-weighting as follows

$$
L_{AE} = 10\log_{10}\left[\frac{\Delta t}{t_0}(10^{L_{A1}/10} + 10^{L_{A1}}/10 + \cdots + 10^{L_{An}/10})\right](dB) \quad (2.9)
$$

where the symbols are the same as in Equation (2.6).

3 *Day-night equivalent level.* This is defined and proposed by the EPA (Environmental Protection Agency) in the USA. It is an average taken

Table 2.6 Permissible values for L_{dn} and L_{Aeq24h}

Purpose	Standard	Application
Audibility protection	$L_{Aeq,24h} \leq 70$	Whole areas
Outdoor activity	$L_{dn} \leq 55$	Residential areas and quiet places
	$L_{Aeq,24h} \leq 55$	School courtyard, public parks and allied spaces
Indoor activity	$L_{dn} \leq 45$	Inside residences
	$L_{Aeq24h} \leq 45$	Schools except residences areas

over 24 hours with a penalty of 10 dB for night-time noise level to take into account the greater annoyance during night as follows

$$L_{dn} = 10 \log_{10} \frac{1}{24}[15 \times 10^{L_d/10} + 9 \times 10^{(L_n+10)/10}]$$

$$L_d = L_{eq}(7.00 \sim 22.00), \quad L_n = (22 \cdot 00 \sim 7 \cdot 00)$$

(2.10)

The recommended criteria are shown in Table 2.6.

4 *Day–evening–night level.* This is defined by the EU (European Union) in a directive concerning environmental noise control in Europe. It is an average taken over 24 hours with a penalty of 5 dB for the evening and 10 dB for the night-time as follows

$$L_{den} = 10 \log_{10} \frac{1}{24} \left[t_d \cdot 10^{L_d/10} + t_e \cdot 10^{(L_e+5)/10} + t_n \cdot 10^{(L_n+10)/10} \right] \text{(dB)}$$

(2.11)

where $L_d = L_{Aeq}$ for the day period (integration time t_d), $L_e = L_{Aeq}$ for the evening (integration time t_e), and $L_n = L_{Aeq}$ for the night (integration time t_n). The time definitions may differ slightly between countries; one example is daytime 7.00–19.00, evening 19.00–22.00, and night 22.00–7.00.

L_{den} is introduced as a measure of noise annoyance and should be applied to all kinds of environmental noise. In the case of road traffic noise, the relation between $L_{Aeq,24h}$ and L_{den} depends on the distribution of the traffic

$$L_{den} = L_{Aeq,24h} + 10 \log_{10} \frac{1}{100} \left[p_d + p_e \sqrt{10} + p_n 10 \right] \text{(dB)}$$

(2.12)

where the percentage of the traffic flow during the daytime, evening and night is p_d, p_e and p_n, respectively.

[**Ex. 2.2**] In a typical example of urban traffic, 78% of the traffic occurs during the daytime (12 hours), 11% in the evening (3 hours), and 11% at night (9 hours). From Equation (2.12), it follows that $L_{den} = L_{Aeq,24h} + 3.5\,dB$.

2.3 Measurement of vibration

A. Human sensitivity to vibration

a. Measurement of equal sensation

Unlike sound, vibration has an extra dimension, that of direction. Furthermore, owing to the complexity of the human sensitivity to vibration, research into it has taken second place to noise. Miwa & Yonekawa (1974) reviewed their experimental work and produced a chart of equal-sensation curves for the human response to vibration (Figure 2.11) as a function of acceleration levels. This figure is similar to the equal loudness curves shown in Figure 1.11. For vibration, however, there are two kinds of curve, one for vertical and the other for the horizontal vibration of the floor on which a person is assumed to be standing or sitting.

b. Evaluation of environmental vibration

ISO 2631 has specified the base curves shown in Figure 2.12, which represent magnitudes of approximately equal human response with respect to

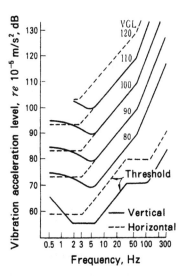

Figure 2.11 Equal-sensation curves for whole body sinusoidal vibration in both vertical and horizontal directions (Miwa & Yonekawa 1974).

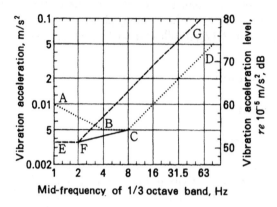

Figure 2.12 Base curves for acceleration in environmental vibration: A–B–C–D for vertical (foot-to-head) direction; E–F–G for horizontal (side-to-side and back-to-chest) direction; E–F–C–D for combined direction, used when the directions of vibration and human occupants vary or are unknown.

human annoyance and complaints about interference with comfort. Satisfactory vibration magnitudes in the human environment should be specified in multiples of the values of the base curves.

Table 2.7 shows the multiplying factors currently used to evaluate the criteria, and Figure 2.13 shows the criteria curves specified by Table 2.7 for the base curve of combined-direction shown in Figure 2.12.

Table 2.7 Multiplying factors applied to the base curves shown in Figure 2.13 to specify satisfactory magnitudes of building vibration (ISO 2631-2)

Place	Time	Continuous or intermittent vibration	Transient vibration excitation with several occurrences per day[a]
Critical working (e.g. medical operation, etc.)	Day Night	1	1
Residential	Day Night	2–4 1.4	30–90 1.4–20
Office	Day Night	4	60–128
Workshop[b]	Day Night	8	90–128

[a]The trade-off between the number of events per day and magnitudes is not well established.
[b]Working places subject to vibration, e.g. drop forges or crushers, may not be included.

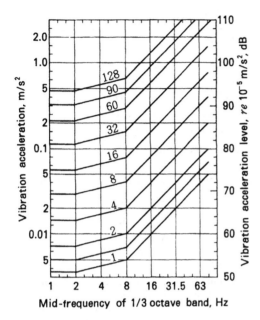

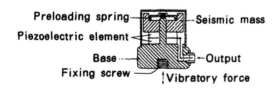

Figure 2.13 Combined-direction criteria curves for environmental vibration, corresponding to the various multiplying factors given in Table 2.7.

B. Measuring instrumentation

a. Construction of the vibration meter

A vibration-measuring instrument works on a similar principle to that of a sound level meter except a vibration pickup replaces the microphone. There are many types of vibration pickup for measuring various kinds of vibration as shown in Table 2.8, such as that produced by noisy machinery or building elements, for example, a wall or floor, that transmit structure-borne sound.

For measurement of environmental vibration, the piezoelectric accelerometer shown in Figure 2.14 is widely used, since it has a very wide frequency and dynamic response with good linearity and, since its output is proportional to acceleration, integration of the output enables velocity or

Figure 2.14 Construction of a piezoelectric accelerometer.

Table 2.8 Vibration pickup: types and characteristics

Mechanism types	Piezoelectric types	Electrodynamic types	Resistant wire strain gauge
Weight	1–600 g	25–4000 g	1–400 g
f_0 resonant frequency	10–180 kHz	0.2–30 Hz	10–2000 Hz
Effective frequency range	a) 1–50 000 Hz	30–4000 Hz	0–700 Hz
	b) 0.1–1000 Hz	0.2–30 Hz	0–8 Hz
Internal impedance	High capacitance	Inductance lower than 1000 Ω	100–400 Ω
Output	Proportional to acceleration at frequencies lower than f_0	Proportional to velocity at frequencies higher than f_0	Proportional to displacement
Purpose	Acceleration, velocity, displacement	Velocity, displacement	Displacement
Merits	Light weight, small size, wide range	High output voltage, cable extension is easy as its impedance is low	Light weight measurement is possible from 0 Hz
Demerits	Preamplifier is required as impedance is high	Easily affected by electromagnetic field	Low output, difficult to amplify

displacement to be measured. The ISO 8041 standard gives directions for vibration-measuring instrumentation for assessing the human response to vibration.

b. Frequency weighting

Human-response vibration-measuring instrumentation should have one or more frequency-weighting characteristics corresponding to the three base curves shown in Figure 2.12 and also an optional flat characteristic to measure unweighted values, not only of vibration but also, for example, of infrasound, using an auxiliary component. The instrument should satisfy the requirements of the specified weighting characteristics and tolerances in a specified frequency range.

c. Time weighting

The detector-indicator should indicate the RMS value using an averaging circuit of 1 s time-constant, i.e. the same characteristics as the S time-weighting of the sound level meter as shown in Table 2.2, because the sensitivity of vibration is reduced when the duration of vibration becomes shorter than 1 s.

d. Vibration measurement and analysis

The vibration pickup should be installed very firmly so as not to wobble or resonate with its own mass coupled to a soft surface. Vibration in three axes should be recorded and frequency analysis performed in 1/3 octaves or with narrow band filters as previously described. In all cases, the effect of ambient vibration should be excluded by using Figure 2.3 and by taking measurements at a number of points. All instruments described in Section 2.1 are also useful as related equipment for measuring vibration.

2.4   Vibration pollution and its distribution

A. Generation of vibration pollution

Annoyance due to vibration is generated by transportation systems, factory machinery and construction work, and is transmitted chiefly through the ground but sometimes through air, causing vibration of buildings and disturbance to the human environment. Not only direct human suffering, but also resonance of building elements, fixtures and furniture producing noise as a secondary effect, results from vibration. In severe cases, physical damage, such as cracking of walls and slipping roof tiles, may occur.

B. Attenuation with distance and prevention of transmission

Unlike noise, ground vibration is propagated by means of longitudinal, transverse and surface waves, all of which have different attenuation characteristics with frequency, direction and amplitude. Moreover, the medium consists of different kinds of soil layers whose transmission characteristics differ locally and include discontinuities due to ground water, etc., thus the transmission mechanism becomes very complicated. In addition, attenuation caused by the medium's internal losses is so variable that exact prediction of the effect of distance may be difficult (Gutowski & Dym 1976).

However, as shown in Figure 2.15, attenuation due to distance seems to be around −3 to −6 dB per doubling of distance, which applies primarily to surface waves, which are largely responsible for the transmission of ground vibration (Shioda 1986). In order to block the transmission path, some obstructions such as ditches and underground walls have been used for some time. In the case of ditches, where the depth is about the same as the wavelength, the amplitude ratio is often reported to be reduced by about 1/10, but there is no reliable formula as yet, which can be used with confidence because of the numerous influencing factors.

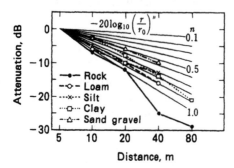

Figure 2.15 Attenuation with distance of ground vertical vibration in different soils (Shioda).

C. Criteria for rating of environmental vibration pollution

The characteristics of vibration pollution may be described as follows:

1 Excluding special cases, vibration does not extend beyond 100 m from the source (in most cases, about 10–20 m).
2 In general, the vertical vibration magnitude is larger than the horizontal one.
3 The vibration frequency commonly lies in the range of 1–90 Hz.

As regards a reliable rating method or criteria for vibration pollution, we can quote only Table 2.7 and Figure 2.13 at present. However, the more detailed evaluation methods for effects of annoyance, especially due to impulsive, intermittent and time-varying vibrations, are currently being researched. There are many difficulties to overcome, because annoyance due to vibration can be changed by its duration and the time period over which events occur; furthermore any startle factor caused by a transient vibration can be reduced by warning signals, announcement and regularity of occurrence and a suitable public relations programme.

2.5 Infrasound and low-frequency noise

Infrasound is defined as sound with frequencies below 20 Hz, which is generally considered the lowest frequency of audible sound. However, if sufficiently powerful, infrasound can be heard or perceived by the human body, but it is not related to a certain pitch. Instead it can be heard as a rumbling or pulsating pressure variation that can be very annoying and sometimes frightening.

Infrasound is generated by large fans, diesel engines, gas turbines, combustion processes in industry and by heavy traffic; also by domestic appliances such as washing machines, waste disposal units, pumps, refrigerators and air-conditioning systems, etc.

Audible sound in the frequency range 20–200 Hz is called low-frequency noise. It may need special attention because noise in this frequency range can be particularly annoying, and it is difficult to evaluate with an objective measurement. The generally used A-weighted sound pressure level fails as a measure of low-frequency noise annoyance.

In the low-frequency region there is such a wide scatter of individual hearing thresholds that a noise audible to one person may not be audible to another. Physiological and psychological factors differ widely between individuals, which is by far the major problem in assessing response to infrasound. The effect can be very distressing for some people who are affected and who are passed back and forth between environmental and medical authorities, both failing to solve their problem (Leventhall 1987).

For the measurement of infrasound, a special G-weighting filter is defined in ISO 7196. It only includes the very low frequencies and has a sharp cut-off at frequencies above 20 Hz (see Table A.5 in the Appendices). A recommended limit for infrasound in dwellings is 85 dB(G).

Several methods have been suggested for the measurement of low frequency noise. A method suggested by (Vercammen 1989, 1992) has proven to be particular reliable and has been adopted in the Danish guidelines for assessment of low-frequency noise. The low-frequency noise is measured in 1/3 octave bands from 10 to 160 Hz and a total level for this frequency range is calculated using the corrections of the A-weighting filter (see Table A.6). Due to the excessive tolerances at low frequencies on the A-weighting filter in the instrumentation standard (IEC 61672-1), the A-weighted level of the low-frequency noise cannot be measured directly. Using this method the recommended limits for low-frequency noise are 5–15 dB lower than the ordinary noise limits. Recommended limits for low-frequency noise in dwellings are 25 dB during daytime and 20 dB for evening and night.

PROBLEMS 2

1 Noise data (band spectrum) analysed by octave-band filters in an office room as shown in Table P.2.1, row X. Obtain the values of dBC, dBA and NCB

Table P.2.1 Octave-band level (dB)

Freq.(Hz)	31.5	63	125	250	500	1k	2k	4k	8k
X	64	68	61	64	57	43	37	29	23
Y	64	68	61	64	61	56	45	34	26

2   In the office space mentioned above, find the reduction value in each octave band necessary to obtain NCB-45. Hence what is the dBA value? Furthermore, when NCB-40 is realised, what is the dBA value?

3   After installation of office machinery in the above room, the noise data changed as shown in Table P.2.1, row Y. Estimate the sound level generated by the machine in each octave band.

4   The results of a noise survey at a roadside are obtained every 5 s as shown in Table P.2.2. Find the statistical values of L_{50}, L_5, L_{10}, L_{90} and L_{95}.

Table P.2.2 Street noise level (dBA) in 5-s intervals

68	71	70	64	65	74	65	68	69	70
73	69	72	69	65	66	69	79	66	67
68	70	69	70	71	73	68	67	70	74
65	66	65	63	63	67	70	68	70	72
69	68	68	71	70	66	65	66	65	67

5   Enumerate the factors related to annoyance of noise and explain the difference between loudness and noisiness.

3 Room acoustics

Acoustic phenomena in a closed room are the main focal problems in architectural acoustics. In this chapter, fundamental theories are described and their practical applications are discussed.

3.1 Sound field in a room

A. Characteristics of the indoor sound field

While the characteristics of the outdoor sound field are rather simple because the sound wave from the source spreads out freely, assuming there is no obstacle, then attenuates with distance, the sound emitted indoors creates a very complicated sound field due to multiple reflections from walls, ceiling and floor. The characteristics of the indoor sound field are as follows. (1) The sound intensity at a receiving point remote from the source is not attenuated as much as in free space even if the distance is large. (2) Reverberation occurs due to the reflected sound arriving after the source has stopped. These two features are quite distinct from the outdoor situation and are very important. Moreover, depending upon the room shape and surface finish, peculiar phenomena like echoes, flutter echoes, etc., which result in a complicated acoustic field, are observed. This is mainly due to the effect of the surrounding walls that determine the room shape. Therefore, the purpose of the study of room acoustics is to control the above described phenomena by means of room boundary conditions such as room shape and finishing materials and to create a satisfactory acoustic environment in the space.

B. Geometrical acoustics and physical acoustics

The science that handles sound energy transmission and diffusion geometrically without considering the physical wave nature of sound is called geometrical acoustics, whereas the science that handles the physical wave nature of a sound wave is called wave acoustics or physical acoustics.

When the wave nature of the sound field in a room is discussed, the wave equation, as described later, has to be solved under the appropriate boundary conditions. It is possible, however, to solve only simple cases. But very complicated problems occur in real rooms surrounded by walls whose shape and characteristics are quite varied. However, in the case of rooms with dimensions that are large compared with the wavelength and with a complexity of walls, the sound field may be analysed rather simply with geometrical acoustics and the wave nature of sound ignored. Thus, most practical problems of room acoustics are handled with the aid of geometrical acoustics although it is necessary to have a proper understanding of wave acoustics for more accurate and effective application of the former.

3.2   Normal mode of vibration in rooms

When wave motion is taken into account in room acoustics, the most important and fundamental characteristic that needs to be understood is the normal mode of vibration of the room. Thus, to begin with, the simplest case of one dimensional space, i.e. the sound field in a closed pipe, is discussed.

A. Normal mode of vibration in a closed pipe

a. Wave equation and its solution

In the case of a closed pipe whose internal diameter is small compared with the wavelength, the sound wave can only propagate longitudinally and not across the pipe, therefore, only one-dimensional plane wave motion need be considered. The wave equation of this free vibration can be derived from the characteristics of the medium as follows (see Section 11.1)

$$\frac{\partial^2 \varphi}{\partial t^2} = c^2 \frac{\partial^2 \varphi}{\partial x^2} \tag{3.1}$$

where φ can be thought of as either the sound pressure p or the particle velocity v because both can be expressed in the same form. Hence, instead of the double process of handling p and v, it is convenient to introduce a new function φ, which is defined as follows

$$\left. \begin{array}{l} v = -\dfrac{\partial \varphi}{\partial x} \\[2mm] p = \rho \dfrac{\partial \varphi}{\partial t} \end{array} \right\} \tag{3.2}$$

where ρ is the density of the medium. The minus gradient in any direction (in this case the x direction) indicates the particle velocity in that direction, hence the reason for calling φ the velocity potential. So, differentiating φ with respect to t or x enables p or v to be derived using Equation (3.2).

When the sound wave can be described by simple harmonic motion, $\varphi \sim e^{j\omega t}$ with angular frequency ω, then Equation (3.1) becomes

$$\left(\frac{d^2}{dx^2} + k^2\right)\varphi = 0 \tag{3.3}$$

where

$$k = \frac{\omega}{c}$$

The general solution of the above equation is

$$\varphi = C_1 e^{j(\omega t - kx)} + C_2 e^{j(\omega t + kx)} \tag{3.4}$$

where C_1 and C_2 are constants.

The first term describes the wave propagating in the $+x$ direction. The second term describes the wave propagating in the $-x$ direction. Alternatively the above solution can be expressed as follows

$$\varphi = (A \cos kx + B \sin kx)\, e^{j\omega t} \tag{3.5}$$

where

$$A = (C_1 + C_2) \quad \text{and} \quad B = -j(C_1 - C_2)$$

b. Natural vibration in a closed pipe

Here, we discuss the case where the pipe is closed at $x = 0$ and l_x by rigid walls. The boundary condition of this space is such that the particle velocity is 0 at $x = 0$ and $x = l_x$. From Equations (3.2) and (3.5) the following is obtained

$$v = -\frac{\partial \varphi}{\partial x} = k(A \sin kx - B \cos kx)e^{j\omega t} \tag{3.6}$$

In order to satisfy the boundary condition $B = 0$ when $x = 0$ and $\sin kl_x = 0$ when $x = l_x$.

Hence,

$$k_m l_x = m\pi, \quad (m = 0, 1, 2, 3, \dots)$$

Therefore

$$k_m = \frac{\omega_m}{c} = \frac{m\pi}{l_x} \tag{3.7}$$

here, $m = 0$ is not our concern as it infers zero vibration. Then, the angular frequency with specific values is expressed as follows

$$\omega_m = \frac{cm\pi}{l_x}$$

and frequencies satisfying the above relation are

$$f_m = \frac{\omega_m}{2\pi} = \frac{cm}{2l_x} \quad \text{(Hz)} \tag{3.8}$$

These are called normal frequencies or natural frequencies characterising natural vibrations, which are also known as normal modes of vibration. Furthermore, at these frequencies the pipe resonates, therefore, they are called resonance frequencies. The wavelengths λ_m for the above case are

$$\lambda_m = \frac{c}{f_m} = \frac{2l_x}{m} \quad \therefore \quad l_x = m\frac{\lambda_m}{2} \tag{3.9}$$

This means that the frequencies where the pipe length is an integral multiple of a half wavelength are natural frequencies. There are an infinite number of normal modes of vibration from $m = 1$ to infinity.

[Ex. 3.1] To obtain the normal mode of vibration $m = 1$ to 3 for a pipe of length 6.8 m. From Equation (3.8) with sound velocity $c = 340 \, \text{m s}^{-1}$

$$f_1 = \frac{340}{2 \cdot 6.8} = 25 \, \text{Hz}, \quad f_2 = 25 \cdot 2 = 50 \, \text{Hz}, \quad f_3 = 25 \cdot 3 = 75 \, \text{Hz}$$

[Ex. 3.2] For the above condition the particle velocity distribution in the pipe whose length is l_x, from Equations (3.6) and (3.7) becomes

$$v = kA \sin\left(\frac{m\pi x}{l_x}\right) e^{j\omega t} \tag{3.10}$$

then the particle velocity distribution is expressed by

$$\sin\left(\frac{m\pi x}{l_x}\right)$$

This is shown by solid lines in Figure 3.1 for $m = 1$ to $m = 3$.

[Ex. 3.3] For the above condition, the sound pressure distribution is derived from Equations (3.2) and (3.5).

$$p = \rho\frac{\partial \varphi}{\partial t} = j\omega\rho A \cos(kx)e^{j\omega t} = j\omega\rho A \cos\left(\frac{m\pi x}{l_x}\right)e^{j\omega t} \tag{3.11}$$

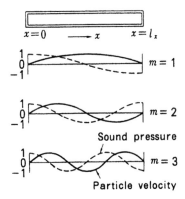

Figure 3.1 Normal mode of vibration in a closed pipe.

Therefore, the pressure distribution is expressed by

$$\cos\left(\frac{m\pi x}{l_x}\right)$$

This is shown by dotted lines in Figure 3.1.

As discussed in the above example, the sound pressure and particle velocity in the pipe in its normal mode of vibration can be determined by position. The wave is found to be a standing wave because its shape does not change.

B. Natural frequency of a rectangular room

In ordinary rooms, the sound wave has to be treated as a three-dimensional field. The velocity potential φ is given as a solution of the three-dimensional wave equation similar to the one for the one-dimensional field as follows

$$\frac{\partial^2 \varphi}{\partial t^2} = c^2 \left(\frac{\partial^2 \varphi}{\partial x^2} + \frac{\partial^2 \varphi}{\partial y^2} + \frac{\partial^2 \varphi}{\partial z^2} \right) \tag{3.12}$$

where c is the sound speed.

Now, we choose a rectangular room with dimensions l_x, l_y and l_z for simplicity, taking the origin of coordinates as the corner of the room, as shown in Figure 3.2. The boundary condition for a rigid wall is that the particle velocity normal to the wall is zero at the wall. The equation can be solved in a similar way to the one-dimensional case because the three variables can be separated. With these conditions the natural frequency is obtained as follows.

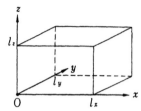

Figure 3.2 Three-dimensional room coordinates.

$$f_n = \frac{c}{2}\sqrt{\left(\frac{n_x}{l_x}\right)^2 + \left(\frac{n_y}{l_y}\right)^2 + \left(\frac{n_z}{l_z}\right)^2} \tag{3.13}$$

where n_x, n_y and n_z are taken as 0, 1, 2, 3, ..., etc. The particle velocity distribution in the room is expressed as a standing wave as follows

$$\sin\left(\frac{n_x \pi x}{l_x}\right) \sin\left(\frac{n_y \pi y}{l_y}\right) \sin\left(\frac{n_z \pi z}{l_z}\right) \tag{3.14}$$

and the pressure amplitude distribution as

$$\cos\left(\frac{n_x \pi x}{l_x}\right) \cos\left(\frac{n_y \pi y}{l_y}\right) \cos\left(\frac{n_z \pi z}{l_z}\right) \tag{3.15}$$

These expressions correspond to Equations (3.8), (3.10) and (3.11) in the one-dimensional case.

Although there are an infinite number of normal modes depending upon arbitrary combination of n_x, n_y and n_z, they are separated into three categories:

1   *Axial mode*: for which two n's are zero, then the waves travel along one axis, parallel to two pairs of walls and are therefore called axial waves.
2   *Tangential mode*: for which one n is zero, the waves are parallel to one pair of parallel walls and are obliquely incident on two other pairs of walls. The waves are called tangential waves.
3   *Oblique mode*: for which no n is zero, then the waves are obliquely incident on all walls and called oblique waves.

[Ex. 3.4] When normal modes (2,0,0), (1,1,0), (2,1,0) exist, the sound pressure distribution can be described in a two-dimensional plane since $n_z = 0$ in Equation (3.15). And the mode (2,0,0) is in one dimension corresponding to the case of $m = 2$ in Figure 3.1. These results are shown by contours of equal sound pressure in Figure 3.3. On the two sides of the

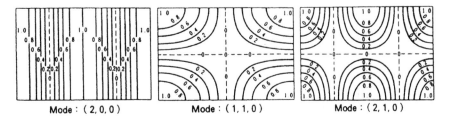

Mode : (2, 0, 0) Mode : (1, 1, 0) Mode : (2, 1, 0)

Figure 3.3 Examples of sound pressure distribution for normal mode of vibrations.

zero contour line of sound pressure, i.e. the node of the standing wave, the sign is reversed as shown in Figure 3.1. For the sound pressure distribution, however, only the absolute values are given since the phase is of no interest.

Some interesting characteristics are illustrated by this example as follows: (1) the sound pressure is a maximum at the room corner for any mode; and (2) when any one of n_x, n_y and n_z is an odd integer, the sound pressure becomes 0 at the centre of the room and so on.

C. Number of normal modes and their distribution

Although the number of normal modes are infinite, when Equation (3.13) is rearranged as follows

$$f_n = \sqrt{\left(\frac{cn_x}{2l_x}\right)^2 + \left(\frac{cn_y}{2l_y}\right)^2 + \left(\frac{cn_z}{2l_z}\right)^2}$$

(3.16)

f_n is found to correspond to the distance between the origin of the rectangular coordinates and the point given by coordinates

$$\left(\frac{cn_x}{2l_x}, \frac{cn_y}{2l_y}, \frac{cn_z}{2l_z}\right)$$

Thus, if a three-dimensional lattice spaced at $c/2l_x$, $c/2l_y$ and $c/2l_z$ along the axes f_x, f_y and f_z, respectively, as shown in Figure 3.4 is formed (called frequency space) then every node of the lattice in the frequency space corresponds to a normal mode, and the number of normal modes is equal to the number of lattice points. Hence, the total number of normal modes less than f is equal to the total number of points within the first octant of the sphere of radius f, having its centre at the origin of coordinates. Therefore, when f increases, the number of points (normal modes) rapidly increases. In this context, f_x, f_y and f_z express the axial modes along the respective axes increasing in proportion to f, while the lattice points located in the plane

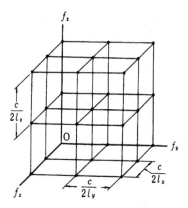

Figure 3.4 Frequency-space lattice showing the normal modes of vibration of a rectangular room.

including two axes express two-dimensional tangential modes increasing in proportion to f^2 and all other lattice points express three-dimensional oblique modes increasing in proportion to f^3.

The volume of the first octant of the sphere with radius f is $(4\pi f^3/3)/8 = \pi f^3/6$. Each mode occupies a volume $c^3/(8\,l_x\,l_y\,l_z) = c^3/(8\,V)$. So, the number of oblique modes below f is approximately

$$N \cong \frac{\pi f^3}{6}\frac{8\,V}{c^3} = \frac{4\pi\,V}{3c^3}f^3 \tag{3.17}$$

Furthermore, if we consider the density of normal modes at frequency f, axial modes are constant regardless of f while tangential modes are almost proportional to the arc length of the circle whose radius is f, therefore the number of modes is proportional to f. Oblique modes are proportional to the surface area of the sphere and thus proportional to f^2. In other words, as far as the distribution of natural frequencies is concerned, the higher the frequency, the larger their density and since the number of oblique modes is largest, the density is considered to be proportional to the square of the frequency. From Equation (3.17) it follows that at the frequency f the modal density is approximately

$$\frac{dN}{df} \cong 4\pi\frac{V}{c^3}f^2 \tag{3.18}$$

D. Room shape and natural frequency distribution

Using Equation (3.13), and depending on the particular choice of (n_x, n_y, n_z), it is possible to obtain the same value for the frequency f_n with more than

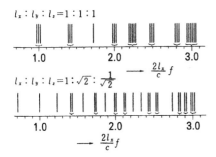

$l_x : l_y : l_z = 1 : 1 : 1$

1.0   2.0   3.0

$\longrightarrow \frac{2l_x}{c} f$

$l_x : l_y : l_z = 1 : \sqrt{2} : \frac{1}{\sqrt{2}}$

1.0   2.0   3.0

$\longrightarrow \frac{2l_x}{c} f$

Figure 3.5 Distribution of normal modes of vibration in two rectangular rooms.

one combination of (n_x, n_y, n_z). These normal modes are then referred to as degenerate. Several lattice points exist on the same sphere, whose centre is at the origin of coordinates in the frequency space shown in Figure 3.4.

The degeneration of normal modes means that their non-uniform distribution makes the acoustic condition of the room undesirable and is related to the dimensional ratio $l_x : l_y : l_z$. The natural frequency distribution in two rooms whose side length ratios are 1:1:1 and $1/\sqrt{2} : 1 : \sqrt{2}$ is shown in Figure 3.5. The normal mode degeneration of a regular cube is very distinct.

In general, when the ratio of the room's length, width to height, is integrally related, e.g. 1:2:4, degeneration is emphasised and, therefore, must be avoided.

The room shape can be used to eliminate degeneration and to create a uniform distribution of normal modes by having oblique boundary planes instead of parallel walls. In such an irregularly shaped room all natural frequencies may be of the oblique mode type, favourable for a uniform sound decay process. This technique is often applied in reverberation rooms where the requirement is for a diffuse sound field.

E. Transmission characteristics

The natural frequency distribution can be obtained by measuring the transmission characteristics of the room. While generating a pure tone of constant intensity with a loudspeaker at one corner of a room and sweeping the frequency, the record of sound pressure level measured at another corner shows the transmission characteristics between the two points. Figure 3.6 illustrates this.

In general, a sharp peak can be found at the normal mode frequencies of the room. Although in the low-frequency range there are fewer normal modes yielding fewer resonance peaks they are, nevertheless, quite distinct. At higher frequencies there are many which are closer together, even overlapping and tending to produce a uniform transmission. From the above

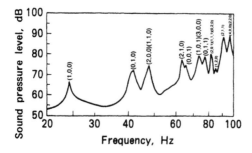

Figure 3.6 Transmission characteristics of a rectangular room of 4.8 · 7.7 · 2.8 m (Knudsen & Harris 1950).

it can be seen that it is desirable to avoid degenerate normal modes and to have a uniform distribution of modes with spacing as nearly equal as possible at all frequencies.

The curve shows various profiles depending upon the positions of the sound source and receiver. All normal modes of vibration of the room can be observed when the sound source is located at a corner and the receiver is in the other corner along a diagonal of the room (cf. [Ex. 3.4]).

F. Effect of absorption of walls

In order to simplify the above discussion we assumed the surrounding walls to be rigid and their surfaces completely reflective. However, they also possess some sound absorption. Therefore, the normal modes of a room generally tend to shift towards lower frequencies, and the rugged profile of the transmission characteristics is flattened. If flat transmission characteristics are desired, it is necessary to select appropriate absorbing materials and construction effective for particular normal modes (see Section 4.5).

3.3   Reverberation time

A. Assumption of diffuse sound field

In geometrical acoustics the sound field in a room is assumed to be completely diffuse. This means: (1) the acoustic energy is uniformly distributed throughout the entire room; and (2) at any point the sound propagation is uniform in all directions.

When the room dimensions become large, the normal modes in the low-frequency range are sparsely distributed but their frequencies are below the range of audibility, while many normal modes build up in the audible-frequency range. It is almost impossible to treat so many natural frequencies as individual wave motions, therefore a statistical approach becomes necessary. In other words, wave motion need no longer be considered and geometrical acoustics seems more reasonable. Also the shape of large rooms is often irregular, sound diffusion dominates and the above assumption is satisfactory.

According to this assumption of a completely diffuse field, room acoustics can be treated very simply whereas it is difficult to find solutions using wave theory. Therefore, an effort to diffuse the sound field in a room must be made in order to utilise the geometrical theory.

B. Sound incidence on wall in diffuse sound field

The energy density E in an infinitesimal volume dV at a distance r from an area dS of wall propagates in all directions in a diffuse sound field. Since the effective area dS as seen from the direction of dV is $dS \cos \theta$ (Figure 3.7), the energy element incident on dS is as follows

$$\Delta I \, dS = \frac{dS \cos \theta}{4\pi \, r^2} E \cdot dV$$

To calculate the total energy E_i incident on dS in 1 s, the sound energy that travels a distance c is integrated for all dV within the half sphere of radius c with its centre on dS. Using polar coordinates r, θ, φ centred at dS

$$dV = r \, d\theta \cdot dr \cdot r \sin \theta \, d\varphi$$

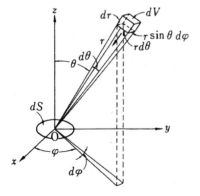

Figure 3.7 Sound incidence from diffuse sound field.

Therefore

$$\Delta I \, dS = \frac{E \, dS}{4\pi} \cos\theta \, \sin\theta \, d\theta \, d\varphi \, dr$$

$$\therefore \quad I \, dS = \frac{E \, dS}{4\pi} \int_0^{\pi/2} \cos\theta \, \sin\theta \, d\theta \int_0^{2\pi} d\varphi \int_0^c dr = \frac{c}{4} E \, dS \qquad (3.19)$$

Hence, the acoustic energy incident in 1 s on a unit area (1 m²) of the surrounding wall, i.e. the intensity I, in the diffuse sound field of energy density E is expressed as follows

$$I = \frac{c}{4} E \qquad (3.20)$$

where c is the sound speed. This is an important feature of a diffuse sound field.

We should remember the above and compare it with the intensity of a plane wave at normal incidence on a wall which is, from Equation (1.16)

$$I = cE$$

C. Reverberation theory

The reverberation time (RT) indicated by Equation (1.36) in Chapter 1 is the most fundamental concept in geometrical acoustics for evaluating the sound field in a room. The theory is based on the assumption of a diffuse sound field in a room, thus, regardless of location and effectiveness of absorbing materials and of the sound source and measuring points, the reverberation time has the same value.

a. Derivation of Sabine's reverberation formula

The total surface area of the surrounding walls of a room is S and the energy density of the diffuse sound field is E. The energy incident on all walls is $cES/4$ from Equation (3.20) and the absorbed energy is $cES\bar{\alpha}/4$ where $\bar{\alpha}$ is the average absorption coefficient of all walls. If the sound source emits a sound power W (watt), the fundamental equation of the total energy in the room is

$$V\frac{dE}{dt} = W - \frac{cEA}{4} \qquad (3.21)$$

where $A = S\bar{\alpha}$ is the room absorption as given in Equation (1.37). With $E = 0$ at $t = 0$ the solution becomes the growth formula for sound in a room

$$E = (4W/cA) \left[1 - e^{-(cA/4V)t}\right] \qquad (3.22)$$

The steady-state value E_0 will be reached when $t \to \infty$ and is given by

$$E_0 = 4W/cA \tag{3.23}$$

Now, when the sound source stops and the sound starts to decay, the total energy reduction in the room is obtained by substituting $E = E_0$, $W = 0$ at $t = 0$ in the fundamental Equation (3.19)

$$E = E_0 \, e^{-(cA/4V)t} \tag{3.24}$$

This is the decay formula. Hence the decay rate is

$$D = 10 \log_{10} e^{(cA/4V)} \; (\mathrm{dB\,s^{-1}}) \tag{3.25}$$

The reverberation time T is the time required for the sound to decay by 60 dB, therefore

$$T = \frac{60}{D} = \frac{6 \cdot 4V}{cA \log_{10} e}$$
$$\therefore \quad T = K\frac{V}{A} = \frac{55.3\,V}{cA} \tag{3.26}$$

Thus, Sabine's reverberation formula (1.36) is derived.
In the above equation

$$K = \frac{24}{c \log_{10} e} \cong \frac{55.3}{c} \quad (\mathrm{s\,m^{-1}}) \tag{3.27}$$

where c is the sound speed as given by Equation (1.7). K does, however, vary with temperature as shown in Figure 3.8.

b. Eyring's reverberation formula and the mean free path

While Sabine's formula is applicable to live rooms with low absorption and hence long reverberation times, and agrees well with experimental results,

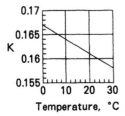

Figure 3.8 Value of K vs air temperature in the reverberation formula.

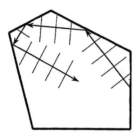

Figure 3.9 A plane wave travelling as a ray from wall to wall in a room.

the formula tends to give values which are too large for dead rooms with much absorption.

For instance, when $\bar{\alpha} \to 1$, i.e. perfect absorption like in an anechoic room, it follows that $T \to 0$, while Equation (3.26) yields a certain finite value since $A = S$. Thus, C.F. Eyring (1931) derived a new formula correcting this defect.

In order to derive a new reverberation formula, the sound decay is studied by following a plane wave travelling as a ray from wall to wall, see Figure 3.9. The energy of the wave is gradually decreased due to absorption at the surfaces, all of which are assumed to have the mean absorption coefficient $\bar{\alpha}$. The ray representing a plane wave may start in any direction and it is assumed that the decay of energy in the ray is representative for the decay of energy in the room. The room may have any shape.

By each reflection the energy is reduced by a factor $(1 - \bar{\alpha})$. The initial energy density is E_0 and after n reflections the energy density is

$$E_n = E_0 \cdot (1 - \bar{\alpha})^n = E_0 \cdot e^{n \cdot \ln(1 - \bar{\alpha})} \tag{3.28}$$

The distance of the ray from one reflection to the next is l_i and the total distance travelled by the ray up to the time t is

$$\sum_i l_i = c \cdot t = n \cdot l_m$$

where l_m is the mean free path. So, the energy density at the time t is

$$E(t) = E_0 \cdot e^{\frac{c}{l_m} \cdot \ln(1 - \bar{\alpha}) \cdot t} \tag{3.29}$$

When the energy density has dropped to 10^{-6} of the initial value, the time t is, by definition, the reverberation time T

$$10^{-6} = e^{\frac{c}{l_m} \cdot \ln(1 - \bar{\alpha}) \cdot T} \Rightarrow -6 \cdot \ln(10) = \frac{c}{l_m} \cdot \ln(1 - \bar{\alpha}) \cdot T$$

This leads to a pair of general reverberation formulas expressed through the mean free path

$$T = \frac{13.8 \cdot l_m}{-c \cdot \ln(1 - \overline{\alpha})} \approx \frac{13.8 \cdot l_m}{c \cdot \overline{\alpha}} \qquad (3.30)$$

The last approximation is valid if $\overline{\alpha} < 0.3$, i.e. only in rather reverberant rooms. The approximation comes from

$$-\ln(1 - \overline{\alpha}) = \ln\left(\frac{1}{1 - \overline{\alpha}}\right) = \overline{\alpha} + \frac{\overline{\alpha}^2}{2} + \frac{\overline{\alpha}^3}{3} + \cdots$$

With the assumption that all directions of sound propagation appear with the same probability, it has been shown by Kosten (1960) that the mean free path in a three-dimensional room is

$$l_m = \frac{4V}{S} \quad \text{(3-dimensional)} \qquad (3.31)$$

where V is the volume and S is the total surface area.

Insertion of (3.31) in the last approximate part of (3.30) gives again the Sabine formula, whereas insertion in the first part of (3.30) leads to Eyring's formula for the reverberation time in a room

$$T = \frac{55.3 \cdot V}{-c \cdot S \cdot \ln(1 - \overline{\alpha})} \qquad (3.32)$$

In a reverberant room ($\overline{\alpha} < 0.3$) it gives approximately the same result as Sabine's formula, but in highly absorbing rooms Eyring's formula is theoretically more correct. In practice the absorption coefficients are not the same for all surfaces, and if surface i has area S_i and absorption coefficient α_i, the mean absorption coefficient is calculated as

$$\overline{\alpha} = \frac{1}{S} \cdot \sum_i S_i \alpha_i \qquad (3.33)$$

It is seen that in the extreme case of an anechoic room ($\overline{\alpha} = 1$) Eyring's formula gives correctly a reverberation time $T \to 0$.

c. Reverberation formula in one- and two-dimensional rooms

Similarly to Equation (3.31), it can be shown that the mean free path in a two-dimensional room is

$$l_m = \frac{\pi S}{U} \quad \text{(2-dimensional)} \qquad (3.34)$$

where S is the area and U is the perimeter. For instance this could be the narrow air space in a double wall, or it could represent the reverberation between reflecting walls in a large, flat room with highly absorbing floor and ceiling.

The one-dimensional case is simply a sound wave travelling back and forth between two parallel surfaces and the mean free path is equal to the distance between the walls, $l = l_m$.

The reverberation time in cases of one- or two-dimensional sound propagation can be calculated from Equation (3.30) by using the appropriate mean free path. In a rectangular room it is often the case that the three-dimensional modes have shorter reverberation times than two-dimensional modes, and the one-dimensional modes have the longest reverberation times, leading to decay curves that do not follow a straight line, as will be discussed later.

d. Reverberation formula including air absorption

A sound wave travelling through the air is attenuated by a factor m, which depends on the temperature and the relative humidity of the air, see Figure 3.10. The unit of the air attenuation factor is m^{-1}. If a plane wave whose intensity I_0 travels x metres the intensity becomes

$$I = I_0 \, e^{-mx} \tag{3.35}$$

If this attenuation is included in Equation (3.29) and putting x equal to the distance ct, which the sound travels, the energy density at the time t is

$$E(t) = E_0 \cdot e^{\frac{c}{l_m} \cdot \ln(1-\bar{\alpha}) \cdot t} \cdot e^{-mct} = E_0 \cdot e^{\frac{ct}{l_m} \cdot \left(\ln(1-\bar{\alpha}) - m \cdot l_m \right)} \tag{3.36}$$

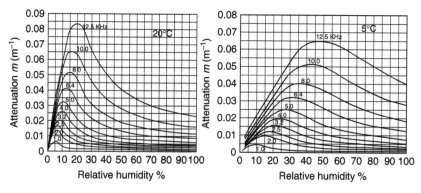

Figure 3.10 Attenuation coefficient due to air absorption (C.M. Harris).

The general reverberation formula (3.30) then becomes

$$T = \frac{13.8 \cdot l_m}{c\left(-\ln\left(1 - \bar{\alpha}\right) + m \cdot l_m\right)} \approx \frac{13.8 \cdot l_m}{c\left(\bar{\alpha} + m \cdot l_m\right)} \tag{3.37}$$

In the most common three-dimensional case, the Sabine formula with air absorption included is

$$T = \frac{55.3 \cdot V}{c\left(\sum S_i \alpha_i + 4\,m V\right)} \tag{3.38}$$

Thus the contribution of air absorption to the equivalent absorption area is $4mV$ (m^2). Some values of the air attenuation factor m are given in Table 3.1.

D. Validity of reverberation formulas

All the reverberation time formulas described above are based on the assumption of a diffuse sound field in a room. Since the decay curve can be expressed in terms of an exponential function, it becomes a straight line on the dB scale. However, when this assumption is not satisfied, the decay curve does not follow a straight line and the reverberation time is difficult to define. For example, when floor and ceiling are highly absorptive but walls reflective, the reflected sounds in the vertical direction decay rapidly while the reflected sounds in the horizontal direction remain repeating reflections with slow decay, thus the decay curve bends as shown in Figure 3.11, and the measured reverberation time, for 60 dB decay, is longer than the values calculated by any formula. Similar phenomena are observed in long tunnels, corridors and large rooms with low ceilings or coupled spaces such as an audience space and stage area with different decay rates where the sound field does not act as a common diffuse space (see Section 11.8).

Table 3.1 Examples of the air attenuation factor m (m^{-1}) at a temperature of 20°C

Relative humidity (%)	Frequency			
	1 kHz	*2 kHz*	*4 kHz*	*8 kHz*
40	0.0011	0.0026	0.0072	0.0237
50	0.0010	0.0024	0.0061	0.0192
60	0.0009	0.0023	0.0056	0.0162
70	0.0009	0.0021	0.0053	0.0143
80	0.0008	0.0020	0.0051	0.0133

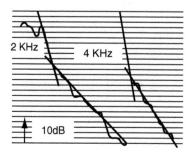

Figure 3.11 Bending in decay curves.

In such cases the calculation of reverberation formulas is no longer useful and the measured reverberation time has little meaning. Instead, it is advisable to show two different decay rates in dB s^{-1} for these cases.

3.4   Sound energy distribution in rooms

When a pure tone from a sound source is generated in a room, the sound field is non-uniform due to standing waves produced by interference with reflected sounds. The sound pressure level distribution shows distinct peaks and troughs as shown by the solid line in Figure 3.12 from measurements at various distances from the sound source. The peak–trough pattern varies with frequency. However, in the case of white noise, which includes all frequencies, the sound pressure attenuates smoothly as shown by the dotted line in the figure. Many of the peaks and troughs made by an infinite number of frequencies overlap each other and smooth out the irregularity of the distribution. Consequently, we need not consider the wave motion.

A. *Non-directional sound source*

The energy density under steady-state conditions from a sound source of power W is given by Equation (3.23), which is the average value for the entire room. However, in reality this density must vary depending on the

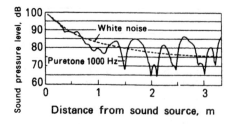

Figure 3.12 Sound pressure level distribution in a room (Knudsen & Harris 1950).

distance from the source. To obtain this distribution, first the energy density due to the direct sound E_d must be obtained. Assuming the sound source is non-directional and the distance to the receiving point is r,

$$E_d = \frac{W}{4\pi r^2 c} \tag{3.39}$$

After this direct sound is reflected at the wall, the energy is assumed to be uniformly distributed in the room as the diffused sound. The energy density is E_s, from which the total diffused sound energy $E_s V$ in the room whose volume is V loses an amount $E_s V\bar{\alpha}$ at every reflection, therefore, every second $E_s V\bar{\alpha}cS/4V$ will be lost due to $cS/4V$ reflections in a second. On the other hand, the energy supply becomes $W(1-\bar{\alpha})$ after the first reflection from the wall. Since this energy is equal to the lost energy in the steady state,

$$E_s = \frac{4W}{cS\bar{\alpha}}(1-\bar{\alpha}) \tag{3.40}$$

Hence,

$$E_0 = E_d + E_s = \frac{W}{c}\left(\frac{1}{4\pi r^2} + \frac{4(1-\bar{\alpha})}{S\bar{\alpha}}\right) \tag{3.41}$$

where $S\bar{\alpha}/(1-\bar{\alpha}) = R$, is sometimes called the room constant.

The relationship between the energy density and the sound pressure is shown by Equation (1.16) as follows:

$$E = \frac{p^2}{\rho c^2} \tag{3.42}$$

Substituting this relation into Equation (3.41), the sound pressure level L_p at the point under normal conditions reduces to

$$L_p = L_W + 10\log_{10}\left(\frac{1}{4\pi r^2} + \frac{4}{R}\right) \quad (\text{dB}) \tag{3.43}$$

where L_W is the power level defined as

$$L_W = 10\log_{10}\frac{W}{10^{-12}\,\text{watt}} \quad (\text{dB})$$

The second term expresses the distribution due to distance from the source as shown in Figure 3.13 for various values of R. When $R \to \infty$, this means complete absorption, which corresponds to open air.

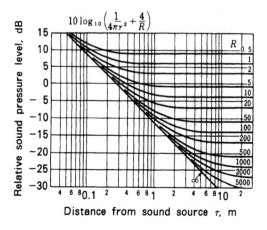

Figure 3.13 Calculation diagram of diffuse sound field in rooms.

B. Directional sound source

When the sound source has a directivity factor Q (see Section 11.13) to the sound receiving points, the direct sound energy density is

$$E_d = \frac{Q\,W}{4\pi\,r^2 c} \tag{3.44}$$

Therefore, in place of Equation (3.41) the energy density is

$$E_0 = \frac{W}{c}\left(\frac{Q}{4\pi\,r^2} + \frac{4(1-\bar{\alpha})}{S\bar{\alpha}}\right) \tag{3.45}$$

The sound pressure level, similar to Equation (3.43) can be expressed as follows

$$L_p = L_W + 10\log_{10}\left(\frac{Q}{4\pi\,r^2} + \frac{4}{R}\right) \quad \text{(dB)} \tag{3.46}$$

In order to carry out further detailed analysis, not only the direct sound E_d but the first reflected sounds are extracted as significant from within the diffused sound, and are counted and superimposed as follows

$$E_0 = E_d + \sum ER_1 + E_{s'} \tag{3.47}$$

where $E_{s'}$ is the diffused energy density due to the second and following reflections. The second term is the total summation of the first reflected sounds from ceiling and walls.

However, for the above assumption of mirror image reflection, the reflecting plane must be sufficiently large compared with the wavelength. If the wall is irregularly modelled or curved, the reflection characteristics need further investigation.

3.5 Echoes and other singular phenomena

A. Echoes

After the direct sound is heard, if a reflected sound is heard separately, it is called an echo, which must be distinguished from reverberation. When echoes occur, speech articulation decreases considerably since speech sounds consist of successive short sounds, and the performance of music becomes difficult because the rhythm is no longer easy to follow. Thus, echoes damage room acoustics quite seriously. However, if the time delay is short, the reflected sound whose intensity may be even 10 dB higher than the direct sound will not be heard separately but will effectively reinforce the direct sound, which is all to the good. Generally, a reflected sound delayed by more than 30–50 ms after the direct sound is recognised as an echo.

The relationship between the time delay of reflected sounds after a direct sound and the relative intensity level and their influence on the degree of damage produced is contained in an index called percentage disturbance proposed by Bolt & Doak (1950), as shown in Figure 3.14. The figure shows what percentage of listeners are disturbed (see Lit. B13). However, the concept did not include the effect of receiving direction nor frequency spectrum change, which add major complications – still further research is required.

B. Flutter echo

When a pair of parallel walls or a ceiling and floor are made of rigid materials, a flutter echo can often occur. A single impulsive sound, such as a hand

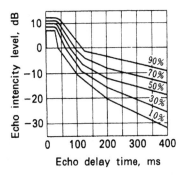

Figure 3.14 Percentage disturbance from echoes (Bolt & Doak 1950).

clap or footstep, produces a multiple echo and is heard as a very peculiar tone such as 'put put put...' or 'purururu...'. This kind of echo consists of multiple repetitions. The Honchido of Toshogu shrine in Nikko, Japan, is famous for this phenomenon of flutter echo, produced by clapping, which sounds as if the painted dragon on the ceiling is neighing, so people call it the neighing dragon.

C. Colouration

When the time delay of reflected sounds from a direct sound is short, i.e. from several to a few 10 ms, the timbre of the sound is changed, to some extent, by phase interference. Sometimes this phenomenon is called colouration.

D. Whispering gallery

If a rigid wall has a large concave surface, a sound moves along the surface with many repeated reflections at grazing incidence, especially at high frequencies. A whisper directed to the surface can be heard distinctly at a distance as great as 60 m. The gallery of the main dome of St Paul's Cathedral in London is famous for this phenomenon.

E. Sound focus and dead spot

As with light, when a sound wave is reflected by a concave surface that is large compared with the wavelength, it concentrates the sound on a spot where the sound pressure rises excessively. This is called a sound focus, which makes the sound field distribution irregular in the room. This means it also produces particular spots at other localities where the sounds are weak and inaudible, called dead spots.

Since the above described phenomena show that the sound fields are not diffuse, those reverberation formulas and sound field calculations based on the assumption of a diffuse field become invalid. Therefore such singular conditions must be avoided as much as possible for good room acoustics. This leads to the necessity of a substantial investigation of room shapes and absorption treatment in designing rooms for good acoustics.

3.6   Measurement and evaluation of room acoustics

The purpose of measuring the acoustics of a room is to determine the sound field accurately and, hence, to evaluate it. The evaluation of the sound field means a psychological subjective judgement based on the use of real ears.

The measurement method necessary for evaluation can only be precisely determined when it is understood what particular physical condition of the acoustics of the room corresponds to what psychological effect is produced.

In this regard there has been much research, though any conclusive views are not yet recognised. In what follows the authors present the state of the art.

A. Sound field at a receiving point in a room

After a direct sound has reached a receiving point, it is then followed by reflected sounds from surrounding boundaries (i.e. walls and ceiling, etc.). Further successive reflected sound waves produce reverberation. Therefore, when an impulse is emitted from a sound source, time-sequential signals are observed as shown in Figure 3.15, which is referred to as an impulse response. Applying a Fourier transformation (see Section 11.3), the transfer function from the sound source to the receiver is obtained, which expresses the transmission frequency characteristics between the points.

Such an evaluation method based on measurement of time-sequential signals was initiated by Sabine (see Lit. B2), who measured the reverberation time with a set of organ pipes and a stop watch. Even now reverberation measurement is still the most substantial topic in the general subject of room acoustic measurements.

[Ex. 3.5] Sabine's measurement method for the reverberation time took advantage of the fact that the steady-state energy density in a room is proportional to the sound power of the sound sources, Equation (3.23). Thus, if he listened to the decay of the sound after stopping four identical organ pipes, the initial energy would be four times higher than in the case of one organ pipe, i.e. the initial sound pressure level would be 6 dB higher. Of course, he was able to hear the decaying sound longer from the four organ pipes than from only one organ pipe, and the time difference would be exactly 1/10 of the reverberation time. Thus the measurements could be done only with a stop watch and a set of organ pipes.

B. Measurement of reverberation

a. Fluctuation of decay curve

Since the reverberation time formula was derived on the assumption of a diffuse sound field, the state of decay shows a typical logarithmic form, so

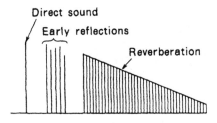

Figure 3.15 Schematic diagram of time-sequential signals at an observing point.

that the reverberation curve is a straight line on the dB scale. The actual measured curve, however, indicates an irregular pattern, so it is necessary to consider the reverberation phenomenon as a wave motion in order to explain it.

A room space is a vibrating system that has many natural frequencies, so that a sound of a particular frequency emitted from a source creates a forced vibration. When the sound source stops and a transient period starts, the source frequency rapidly vanishes, and the vibrational energy shifts to the natural vibrations causing reverberation where the frequencies are close to the source frequency. In the reverberation process, those naturally excited vibrations mutually interfere so that the decay curve has an amplitude fluctuation caused by the beating of different natural frequencies one with another. Each natural vibration has its own decay rate and those of smaller decay rate remain to produce a bending of the decay curve, resulting in a deviation from the logarithmic attenuation.

If a pure tone is used as a source when measuring reverberation, only a few natural vibrations in the reverberation are excited and their interference produces more distinct peaks and troughs, and bending of the decay curve as shown in Figure 3.16(a), therefore, a single decay rate cannot be determined. On the other hand, with a warble-tone or broadband noise, many natural frequencies are excited so that superposition of more interferences and bendings occurs, resulting in a smoothing out of the fluctuation in the decay process, so that the decay curve appears to be a straight line.

At low frequencies, where the number of natural vibrations is small, even where warble-tone or band noise is used, fluctuation or bending may be apparent. At high frequencies the decay curve is a straight line because of the large number and high density of natural vibrations. Therefore, it is clear that the assumption of a diffuse sound field in geometrical acoustics (see Section 3.3.A) is justified.

Two basically different methods are used for measuring the reverberation decay curve; either the interrupted noise method or the integrated squared impulse response method. The latter is based on the finding by Schroeder (1965) that the smooth curve obtained by integrating the squared impulse response is the same decay curve that would be obtained after averaging an

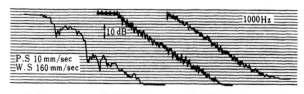

(a) Pure tone   (b) Warble tone   (c) ⅓ Oct. Band noise

Figure 3.16 Reverberation curves recorded by a high-speed level recorder for different sources.

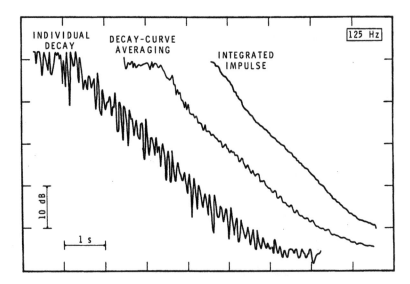

Figure 3.17 Reverberation decay curves obtained with different methods. From left to right: individual decay by interrupted noise; ensemble-averaged decay of 30 measurements with interrupted noise; integrated squared impulse response (Chu). Reused with permission from J. Acoust. Soc. Am.

infinite number of decay curves obtained with the interrupted noise method, see Figure 3.17. The commonly used methods are described in further detail in the international standards ISO 3382-1 and ISO 18233.

b. Choice of sound source

In reverberation measurement as described above, a sound source, which can excite many natural vibrations in the measuring frequency range, is used. The following are the most commonly used sources in practice.

(a) *Band noise*: it is usual to filter a white noise source into an octave of 1/3 octave bands in order to obtain greater sound power around the measuring frequency with a given loudspeaker. It is widely used because it is easier to handle than warble-tone.

(b) *Warble-tone*: the frequency is varied with a modulation frequency a within a bandwidth of $\pm \Delta f$ at the mid-frequency f. Its spectrum is not continuous but has discrete lines with a frequency spacing a mainly inside the bandwidth.

(c) *Short tone or AC-pulse*: speech and music are time-sequential signals consisting of short sounds, by which the room characteristics are often measured in the following ways.

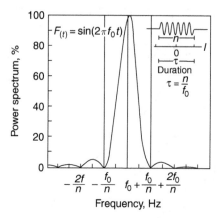

Figure 3.18 Power spectrum of a pure toneburst.

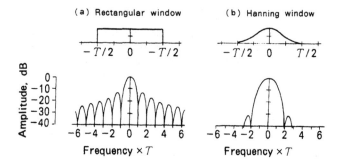

Figure 3.19 Difference of spectra owing to time windows.

1   *Toneburst or pulsed-tone*: these are generated by slicing a pure tone into a few to several tens of waves. The spectrum has its own width around the original frequency f_0 and contains most of the power within the frequency range $\pm f_0/n$, as shown in Figure 3.18. When natural vibrations in the range of $\pm 10\%$ of the measuring frequency are to be excited, 5–6 waves are found to be required. This corresponds to a rectangular time window as shown in Figure 3.19(a), while the other windows can be used to further concentrate the power into the original frequency, as in an example shown in Figure 3.19(b).

2   *Noise-burst*: breaking white noise or band noise into a few to several tens of milliseconds. In the measurement system it is usual to use a band filter that has its mid-frequency at the frequency of interest.

(d) *Shock wave or DC-pulse*: since an ideal impulse has a uniform continuous spectrum over all frequencies (see Section 11.3), this is used in the same way as white noise. In the laboratory, the spark due to an electric discharge from a condenser (8–10 μF and 0.5–10 kV) is often used, on the other hand, because of easy control and repeatability, DC-pulses, which are emitted from a loudspeaker energised with electrical rectangular signals having widths from several to scores of microseconds generated by a signal generator, are often used. However, since the intensity is small, it is necessary to practice synchronous summation for measurement. A signal-pistol shot used as a source has sufficient power so that the SN ratio is high, except in the lowest frequency range, and its ease of use makes it popular in field measurements.

(e) *MLS signal*: the sound signal is made from a mathematical code called a Maximum Length Signal (MLS). The spectrum is broadband like white noise, but by applying an auto-correlation process to the recorded signal, it is possible to derive the impulse response. The signal-to-noise ratio is very good and can be further improved by averaging repeated sequences. The weakness of the method is that the transfer function from source to receiver must be absolutely constant during the measurement; sometimes small air movements or temperature variations can interfere with the measurements. The method cannot tolerate distortion in the loudspeaker. Further information is found in ISO 18233.

(f) *Sine-sweep*: this is a very simple signal, a pure tone with a frequency that increases from very low to very high, covering the whole frequency range of interest. The speed of the sweep can be quick or slow. The most efficient method is to use a very slow sweep and an exponential increase of frequency, i.e. the sweep takes the same time in each octave band. By a deconvolution processing of the measured signal, the impulse response is derived. The signal-to-noise ratio is very good and can be further improved with a longer sweep time. Other advantages are that distortion in the loudspeaker is no problem, and the requirement of time-invariant measurement conditions is less restrictive than with the MLS signal. The demands on the computer power are higher than with the other methods, but this is no problem with the technology today. Further information is found in ISO 18233.

Except for the electrical discharge and the pistol shot, electrical signals described above are amplified and the source sound generated from an omnidirectional loudspeaker. Figure 3.20(a) shows a dodecahedron-type speaker, which has twelve transducers producing sufficient power overall. When a point source is necessary using a DC-pulse, an arrangement with two loudspeakers, as shown in Figure 3.20(b), is used.

c. Sound receiving and observation

For sound reception, an omnidirectional sound pressure sensitive microphone is used. The output may be taken either directly to an amplifier, filter set and a system for displaying decay curves or analysis equipment

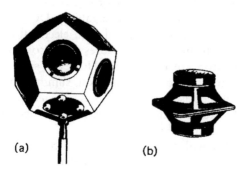

(a)   (b)

Figure 3.20 Omnidirectional speakers for measurement: (a) dodecahedron-type speaker; and (b) confronting speaker.

for deriving the impulse responses; or to a signal recorder for later analysis. For measurements *in situ*, it may be convenient, for example, when using pistol shots, to receive them via a sound level meter and record the decay with a tape-recorder or digital recording device on the spot. The signals can then be reproduced later and analysed in the laboratory. When using this technique it is important to confirm the linearity and the dynamic range of the microphone and the recording system, i.e. to be sure that the same signal as the original can be faithfully reproduced.

Although the reverberation time is defined as the time required for the sound decay by 60 dB, this range is not used in practice for the evaluation of the decay. Instead the evaluation range is taken from -5 to -35 dB below the initial level of the decay curve, and, preferably, a linear regression analysis is applied to find the slope of this part of the decay curve. The reverberation time is then found by extrapolation of this line to the full range of 60 dB. Since the evaluation range is 30 dB the result is sometimes denoted T_{30}. A smaller evaluation range from -5 to -25 dB below the initial level is also in common use, and then the result is denoted T_{20}.

(a) *High-speed level recorder*: This is the classical method for reverberation measurement, but is not much in use any more. This apparatus records the varying sound pressure level automatically on a moving roll of paper. The pen's writing speed can be varied arbitrarily up to $1000\,\mathrm{dB\,s^{-1}}$ and the paper speed can also be varied over a wide range. This means that high-speed level recorders can be used for many acoustic measurements. However, because of an automatic balancing mechanism with a moving mass, the peak values of impulse sounds may not be accurately reproduced and the pen speed is limited.

When measuring the reverberation time T, the pen speed α must satisfy the relation $\alpha \geq 120/T$ $(\mathrm{dB\,s^{-1}})$ in order to follow the decay exactly. Therefore, even with the highest speed, measurements with $T < 0.12\,\mathrm{s}$

cannot be achieved. Although the reverberation time is defined as the time required for the sound decay by 60 dB, it is often difficult to observe such a decay range owing to the background noise level, therefore, in practice, the slope of the curve down to around 30 dB is extrapolated to obtain the reverberation time.

(b) *Instruments with digital technology*: Modern acoustic measurement instruments are based on digital technology, and for room acoustic measurements there are several software packages available to be used with a portable computer. The digital technology allows automatic signal processing and analysis of the results, which can be displayed as graphs or tables. However, for reverberation measurements it is also important to have the possibility to view the decay curve, and the operator should have the necessary theoretical background to evaluate whether the results are reliable.

d. Measurement method

Strictly speaking, the assumption of a completely diffuse sound field may not be satisfied, therefore, as many measuring points as possible should be used. At least two source positions and two microphone positions should be used. The distance between measuring points should be at least half a wavelength at the lowest frequency of interest in order to ensure statistically independent measuring points. This may be difficult to achieve at low frequencies in small rooms, but still as many points as possible should be used and the number of measurements should be increased because the fluctuations may be large.

For the interrupted noise method, the measurements should be repeated several times at each point, whereas for the integrated squared impulse response method, one single measurement is sufficient at each point. The average value of all measurements is then used to describe the reverberation time in the room. ISO 3382-2 recommends at least six combinations of source-receiver-positions and at least two repeated measurements at each point. This is for the so-called engineering level of accuracy; for a precision-level measurement, at least twelve source-receiver-positions are required with three repetitions at each point.

The frequency range for reverberation time measurements should at least cover the six octave bands with centre frequencies from 125 to 4000 Hz. However, for a more detailed measurement the 1/3 octave bands from 50 to 10 000 Hz may be applied.

C. Evaluation of reverberation characteristics

Reverberation time is considered as a quantity which can be used to control room acoustics and, so, over many years, many researchers have proposed optimum conditions based on reverberation characteristics.

a. Optimum reverberation time

Reverberation gives sound a rich sonority, so that it is preferable to have a fairly long reverberation time for the performance of music. while for speech and lecturing articulation improves as the reverberation time gets shorter. So, depending upon the particular use of the room, the optimum value of the reverberation time has to be varied.

With a shorter reverberation time, a higher articulation is obtained. On the other hand, when the reverberation time is short the total absorption in the room is large, therefore the sound energy density in the room is low. When the speech level is lowered the percentage articulation (PA) tends to be lowered. These factors can be clearly understood with the aid of Figure 1.19. As a result of two conflicting factors, there is a certain value of reverberation time that may produce the highest value of PA.

Figure 3.21 shows that the peak of the curve indicating the relationship between the reverberation time and PA varies as a function of room volume. If the room becomes larger, the reverberation time is preferably slightly longer (see Lit. B4).

In the past many researchers have raised various proposals concerning the functional relationship between the optimum reverberation time and the room volume, both empirically and theoretically. A representative relationship is shown in Figure 3.22.

Proposals by Beranek (Lit. B13a), Brüel (1951, Lit. B10) and Knudsen and Harris (1950, Lit. Al) are compared in Figure 3.23 and it can be seen that there are fairly wide differences between them, which indicates that the evaluation of the auditory environment is reflected in local and traditional preferences.

b. Frequency characteristics of reverberation time

The human perception of reverberation depends not only on the reverberation time at 500 Hz but also on the variation in reverberation over the entire frequency range. The importance of this perception of reverberation has attracted the attention of many researchers, who have

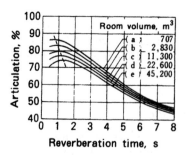

Figure 3.21 Percentage articulation vs reverberation time for different room volumes (Knudsen 1932, Lit. B4).

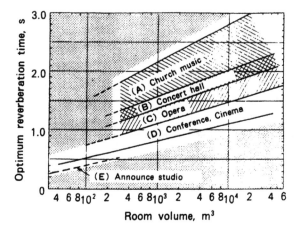

Figure 3.22 Optimum reverberation time at 500 Hz for different types of rooms (Maekawa 1990).

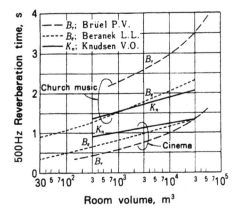

Figure 3.23 Comparison of optimum RT by three authorities.

made various recommendations regarding optimum characteristics as shown in Figure 3.24.

From the point of view of auditory sensation, in order for the loudness level to attenuate uniformly over the entire frequency range, at both low and high frequencies the reverberation time must be long. However, it becomes short at high frequencies because of air absorption, a condition under which sounds are normally heard. Therefore, if the reverberation at high frequencies is made long, the acoustic environment feels unnatural. Also in order not to reduce the percentage articulation of speech, a shorter reverberation at low frequencies rather than mid frequencies is preferred. This is particularly true when microphones for amplification and broadcasting are used.

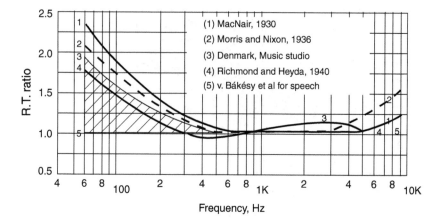

Figure 3.24 Recommended reverberation-frequency characteristics curves, generally, curve 4: for music, and 5: for speech, and shaded area: tolerance for multi-purpose.

Therefore, the frequency characteristics of reverberation must be flat over the entire frequency range. It may be noted that for music, as shown in Figure 3.24, reverberation should be slightly raised within the shaded area. Although the permissible range might be fairly wide, as seen in Figure 3.23, sharp variations in reverberation with frequency must be avoided because of the possibility of peculiar effects on timbre.

c. Indices for assessing the feeling of reverberance

Even in a room that satisfies the requirement of optimum reverberation time, reverberance depends on the hearing location. Even if two rooms have the same volume and same reverberation time, we often experience quite different reactions to their acoustics. These facts show that reverberation time is not enough to describe the acoustics of the room completely, though it is an important factor.

The reverberation time has been determined so far with no reference to many other factors affecting the hearing sensation, such as the distribution of absorbents and the positions of sound source and receiver. Therefore, it is necessary to introduce some indices that will supplement the optimum reverberation time in order to specify the acoustics of the room more satisfactorily. Since many researchers have made various proposals, some representative ones are included in the following:

1   Jordan (Lit. B31) suggested that since the sensation of reverberance seems to correlate well with the early slope of the decay curve, the time

required for a 10 dB-decay multiplied by six might be more useful. This is called EDT (Early Decay Time). In addition to this early 10 dB, the time required for 15 or 20 dB decay has also been proposed by other workers.

2   Generally, a room is called live when a preponderance of reflected sounds produces a reverberant feeling. On the other hand, if direct sounds are dominant and there are too few reflected sounds, it is called dead or dry. Attempts have been made to express liveness as the ratio of the direct sound energy density E_d (Equation (3.39)) to the total energy density E_0 (Equation (3.23)) in the steady state. Cremer called the distance from the source where this ratio becomes unity, the *Hallradius* (German) or reverberation radius. There is another proposal using the diffuse sound E_s (Equation (3.40)) in place of E_0. In either case, when the distance increases from the source, the direct sound gets less, thus the ratio becomes large.

On the other hand, recent psychological measurements show some examples where the reverberant feeling becomes a maximum immediately in front of the stage in a hall. This is due to the time delay of reflected sounds from the rear wall, though they might be considered as echoes, resulting in the reverberant feeling. The problem seems to lie not in the steady state but to be latent in the time-sequential signals.

D. Time-sequential signals: measurement and evaluation

The basis of time-sequential signals is an impulse response, and ideally speaking, it should include all the acoustic information in it. In addition to the direct measurement of DC-pulses, as described previously, the correlation method with white noise as source and the high speed process with M-sequential signals using a digital computer have also been used for analysis. Though it is an approximation, the DC-pulse response is useful for the evaluation of the hearing sensation, and is called the echo time pattern

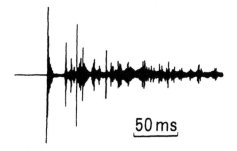

Figure 3.25 Echo time pattern.

(Figure 3.25). The delay time of individual early reflected sounds following the direct sound, corresponds to the transmission distance. The effects of the intensities, densities and the delay times on the sensation are being investigated. When a certain time (>100 or 200 ms) elapses, many reflected sounds overlap to form the reverberation curve. Therefore, the true information is assumed to be included, not in the reverberation time, but in the echo's fine structure of up to 100–200 ms after the direct sound reaches us, thus more detailed analysis can be made.

a. Evaluation on time axis

Kuhl (1957) noticed that the larger the initial time delay gap Δt, between the direct sound and the first reflected sound, the larger the spatial impression that might be obtained, and suggested that a delay of about 30 ms would be desirable (see Lit. B29), while Beranek stated that it would be better if it were less than 20 ms (see Lit. A4).

b. Evaluation using integrated square value of sound pressure

In order to evaluate the pattern of time fluctuation of energy density, Thiele (1953, Lit. B29) introduced an index named *Deutlichkeit* (German) or definition as given in Equation (3.48), which presumes that the square of the instantaneous sound pressure within the first 50 ms is effective in reinforcing the direct sound.

$$D = \frac{\int_0^{50\,\mathrm{ms}} p^2(t)\,\mathrm{d}t}{\int_0^{\infty} p^2(t)\,\mathrm{d}t} \tag{3.48}$$

Thus, it was stated that the larger the value of D the more the distinctness of speech.

Reichardt (1972, Lit. B29) has developed the principle for music by modifying the integration range to 80 ms, separating the initial sounds from the diffused ones as follows,

$$C = 10\log_{10}\frac{\int_0^{80\,\mathrm{ms}} p^2(t)\,\mathrm{d}t}{\int_{80\,\mathrm{ms}}^{\infty} p^2(t)\,\mathrm{d}t} \tag{3.49}$$

which is designated *Klarheitsmass* (German) or clarity.

In addition, there are various proposals taking into account the integration range of the ratio (see Lit. B29a, b).

c. Time weighting of energy method

Cremer proposed an index with time weighting for the evaluation of the decay curve as follows.

$$T_S = \frac{\int_0^\infty t p^2(t)\mathrm{d}t}{\int_0^\infty p^2(t)\mathrm{d}t} \tag{3.50}$$

which is called *Schwerpunktzeit* (German) or point of gravity time and/or centre time. Kürer (1969) has shown that syllable articulation has good correlation with the T_S values (see Lit. B29).

When the sound field in a room is completely diffuse and the reverberation has an exponential decay, the indices of Equations (3.48)–(3.50) have perfect correlations with the reverberation time, so that no new information is provided. Only when the actual sound field is not completely diffuse do the indices provide new meanings.

E. Measurements related to transmission characteristics

It is very important that the sound generated by a sound source is transmitted to the listener with high fidelity. The transmission frequency characteristics measured with pure tone as shown in Figure 3.6, however, are used to detect the natural frequencies but are not suitable for measuring the performance of the room considered as an acoustic transmission system. Methods for providing this information are described below.

a. Transfer function

Applying the Fourier transformation (see Section 11.3) to the impulse response in the time domain enables a transfer function in the frequency domain to be obtained in which the real part expresses the amplitudes and the imaginary part the phases. Recently FFT (Fast Fourier Transform) analysers have become available by which the analogue signals are converted into digital form by an A-D converter, the Fast Fourier transformation carried out and a transfer function displayed almost in real time.

b. Measurement of transmission characteristics with noise signal

For coordination and performance testing of electroacoustic systems, a method of analysing the microphone output in 1/3 octave bands at the receiving point during the steady emission of white noise is used. The white noise has a constant spectrum in which the band level of constant ratio bandwidth tends to rise with a slope of 3 dB oct^{-1} as shown in [Ex. 1.9]. Therefore, if the sound source emits so-called 'pink noise' with a filter which has the reverse slope, the entire measuring system has a flat characteristic. Hence, the measured results immediately indicate the characteristics of the transmission system with a real time analyser.

c. Measurement of MTF (modulation transfer function)

Distortion in speech signals caused by reverberation, echoes and noise in their transmission path to the listener, reduces speech intelligibility. Since speech is composed of a series of pulses whose amplitudes fluctuate sharply on the time axis, reverberation and noise fill these sharp dips, thus distortion occurs. When modelling the above phenomena, using amplitude-modulated noise, the modulation reduction factor m for 100% modulation at the sound source is measured at the observer. Houtgast & Steeneken (1973) propose to measure the m-values at fourteen modulation frequencies of 1/3 octave spaced over 0.63–12.5 Hz, whilst modelling the amplitude fluctuation pattern of speech using seven octave bands of noise from 125 Hz to 8 kHz. The MTF curves are constructed from 98 measured m values as shown in Figure 3.26. An STI (Speech Transmission Index) is proposed using the same calculation method as that employed in deriving the articulation index. Furthermore, in order to simplify the measuring procedure the RASTI (Rapid STI) method is proposed. These methods have

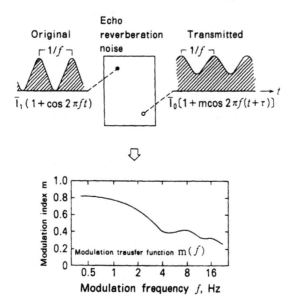

Figure 3.26 Principle of measuring MTF in a room as a function of modulation frequency. Typically, the carrier is 1 octave-band noise (Houtgast & Steeneken 1973).

Table 3.2 Relation between STI and speech intelligibility

STI	Speech intelligibility
0.00–0.30	Bad
0.30–0.45	Poor
0.45–0.60	Fair
0.60–0.75	Good
0.75–1.00	Excellent

This extract is reproduced with the permission of the Danish Standards Foundation. The extract is based on the Danish Standard DS/EN 60268-16:2003. © Danish Standards Foundation.

more merit due to objective and physical measurement compared to a direct measurement of the percentage articulation using human voices and auditory sensation.

The STI measurement method has been standardised in IEC 60268-16. It is often used for the evaluation of loudspeaker systems or sound reinforcement, e.g. in churches with difficult acoustic conditions. For good speech intelligibility, STI values should be as high as possible, see Table 3.2; however, in very reverberant rooms it may be difficult to reach higher values than 0.60 even with sound reinforcement.

Recently, the STI measurements have also been proposed as a measure of speech privacy in open plan offices, in which case values below 0.20 are recommended and values below 0.45 are considered acceptable.

F. Measurement of spatial information and evaluation

a. Measurement of sound pressure distribution

While generating a steady octave-band noise from a sound source (Figure 3.20(a)), the changes in sound pressure level with distance are continuously recorded with a moving microphone as shown in Figure 3.27, where the dotted line indicates the calculated values from Equation (3.46) assuming a completely diffuse field. In practice, a pink-noise source is used, the signal received by the moving microphone is recorded and then analysed in the laboratory, or a direct measurement is performed at several selected points in the hall. In the latter method, if the power level of the source is properly calibrated, the absolute values of the sound pressure levels are obtained in the hall so that they can be compared with values measured in other halls.

The sound strength, G, in a concert hall can be measured using a calibrated omnidirectional sound source, as the sound pressure level relative to that measured at a distance of 10 m from the same sound source in a free field, as described in ISO 3382-1.

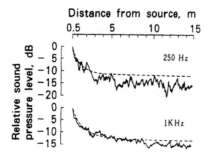

Figure 3.27 Sound pressure level distribution in a room in a steady state.

For the evaluation of the spatial decay of the sound from an omnidirectional sound source, the attenuation of the sound pressure level per distance doubling DL_2, can be measured as described in ISO 14257. The sound pressure level is measured at a number of microphone positions on a line and in different distances from the source. When the sound pressure levels are displayed as a function of the distance on a logarithmic axis, a linear regression line through the measurement points can be calculated. The slope of the line is expressed in dB per distance doubling. In an ideal free field DL_2 should take the value of 6 dB, whereas it should be 0 dB in an ideal diffuse sound field. This measure is used to describe the acoustic conditions in large workrooms, open plan offices, etc., and a value $DL_2 \geq 4\,dB$ is sometimes recommended as acceptable.

b. Evaluation of diffusivity of a sound field

Although geometrical acoustics based on the assumption of a completely diffuse sound field provides a powerful tool for solving practical problems, there might be some problems based on dissatisfaction with the assumption that, for instance, complete diffusion may not be considered the best acoustic condition in all rooms. It is an interesting and unknown problem as to what degree of diffusion exists in an actual room, and how much diffusivity is to be expected in a particular room.

1   *Evaluation with reverberation time and sound pressure distribution*: the irregular bending of the reverberation curve, the scattering of measured values of reverberation time and the deviation from the theoretical value of the sound pressure distribution (Figure 3.27) are considered to express poor diffusion, although evaluation has not been normalised yet.

2   *Evaluation due to correlation of sound pressures between two points*: the correlation R of sound pressures between two points in a completely

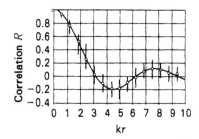

Figure 3.28 Sound pressure correlation values between two points measured in a well-diffused room and the theoretical curve (Cook *et al.* 1955).

diffuse sound field is proved theoretically to be $R = \sin kr/kr$ where r is the distance between the two points and $k = 2\pi/\lambda$. Cook *et al.* (1955) carried out a series of measurements on the correlation between the outputs of two omnidirectional microphones whilst changing the distance r between them in a reverberation room that appeared to be sufficiently diffuse, and observed good agreement with the theoretical curve as shown in Figure 3.28. When the measured values are close to the theoretical ones, diffusion is considered good and the deviation might be considered a measure of that diffusion.

[**Ex. 3.6**] From the result in Figure 3.28 it can be stated that for measurements in a diffuse sound field, two microphone positions should not be closer than half a wavelength; on the other hand, if this is fulfilled the measurement positions with reasonable approximation can be considered as uncorrelated measurement positions.

c. Evaluation due to lateral energy

Although the sound field is a spatial event, nevertheless, the physical behaviour of sound waves at one point has been considered, and an omnidirectional microphone is used for measurements. However, a person hears speech and music with both ears, which produce a binaural effect (see Section 1.12). With this effect one can not only localise the sound source but can gain a spatial impression.

Spatial impression is a very important factor in the evaluation of room acoustics. The greatest contributors to this factor are the lateral reflections. Figure 3.29 shows an experimental result with a single reflection, the effect of a lateral sound, i.e. spatial impression is almost the same within a 10–80 ms delay. But, the more energy in the lateral reflection the greater the spatial effect obtained, as shown in Figure 3.30. This fact is caused by dissimilarity between signals received by the two ears (see Lit. B25).

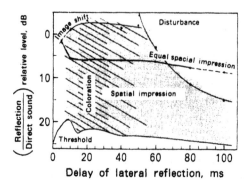

Figure 3.29 Subjective effects of single reflection with lateral angle variable delay and level for music (Barron 1971).

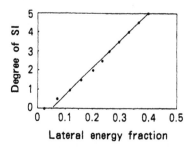

Figure 3.30 Degree of spatial impression vs lateral energy fraction of received sound (Barron & Marshall 1981).

A measurement method using directional microphones has been devised in order to evaluate the binaural effect due to reflected sounds from either the left or right.

In a concert hall the lateral reflections can be measured with a 'figure of 8' directional microphone mounted with the maximum sensibility pointing to the side directions. There are two different aspects of spaciousness that can be evaluated by the lateral reflections. It has been found that the early lateral reflections, i.e. up to a delay of 80 ms after the arrival of the direct sound, correlate with the subjective impression of the size of the sound source, called Apparent Source Width (ASW). The associated measurement parameter is the Early Lateral Energy Fraction (*LF*), which is defined as

$$LF = \frac{\text{early lateral energy } (25-80 \text{ ms})}{\text{early total energy } (5-80 \text{ ms})} \qquad (3.51)$$

The late lateral reflections, i.e. those arriving more than 80 ms after the direct sound, have been found to correlate with the subjective impression of being embedded in sound, called Listener Envelopment (LEV). The associated parameter is the Late Lateral Sound Level (LG), in dB, which is defined as

$$LG = 10 \log_{10} \frac{\text{late lateral energy (80 ms} -\infty)}{\text{free field energy in 10 m}} \quad \text{(dB)} \qquad (3.52)$$

d. Measurement of directional diffusivity

In order to show the intensity of sound incident at a receiving point in any particular direction, measurements are carried out in three dimensions using a rotating unidirectional microphone. The results are presented in the form of 'needles', the length of which are proportional to the intensity in the various directions as shown in Figure 3.31, and which resemble a 'hedgehog'. With a total number N of needles each of whose length is A_i, the mean value is as follows

$$M = \frac{1}{N} \sum_{i=1}^{N} A_i$$

The average deviation from M in every direction is

$$\Delta M = \frac{1}{N} \sum_{i=1}^{N} |A_i - M|$$

Putting $m = \Delta M/M$ and m_0 being its value in free space, Richardson and Meyer (1962, Lit. B17) derived the following expression

$$d = 1 - \frac{m}{m_0} \qquad (3.53)$$

which is called directional diffusivity. In free space $d = 0$. In a completely diffuse sound field, $m = 0$, hence $d = 100\%$.

Plan   Side elevation

Figure 3.31 Hedgehog showing directional diffusivity (Richardson and Meyer 1962, Lit. B17).

In practice, this is measured using a parabolic microphone whose diameter is 1.2 m and narrow band 2000 Hz steady noise. The measured values of d in many halls and studios are within 20–80%, and the larger the room volume the smaller the values of d obtained (see Lit. B17).

As described above, the diffusivity of a sound field may be directly evaluated only in terms of its directional diffusivity. However, it has not been a popular technique largely due to the fact that it is a difficult measurement to make.

e. Closely located four-point microphone method

Yamasaki *et al.* (1989) have recently developed a computer-aided method using four omnidirectional microphones, where one is located at the origin and the other three microphones are located several centimetres away from the origin along three orthogonal axes as shown in Figure 3.32(a). With these four microphones, the impulse responses at a receiving point are measured for the DC-pulses radiated from a source (Figure 3.20(b)), then the arriving sound's direction, intensity and equivalent positions of image source are obtained using signal processing with a digital computer. By this method the intensity of each reflected sound, the direction from which it has come, and the time sequence of the arriving impulses may

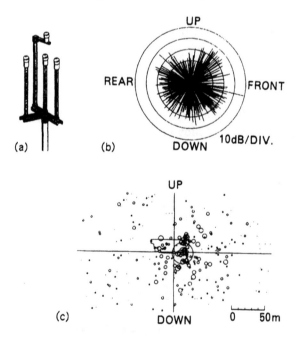

Figure 3.32 A result measured by closely located four-point microphone method (Tachibana *et al.* 1989): (a) four closely located microphones; (b) directivity pattern of received sound; and (c) virtual image source distribution.

be obtained more accurately, therefore all necessary information can be obtained in a very short time. Figure 3.32(b,c) shows one example of a measured result by this method. Further research should be pursued on how to express the data corresponding to the aural sensation from the vast amount of data contained in the four-dimensional information obtained by this technique.

f. Measurement of IACC (interaural cross correlation)

Spatial impression is caused by the dissimilarity of received sounds by two ears, as mentioned above. The dissimilarity is expressed mathematically by the cross-correlation function (see Section 11.3). The lower the value of cross correlation, the greater the dissimilarity (Damaske 1967/68). Consequently, Ando (Lit. B33) defined and adopted the IACC (see Section 11.3.D) as an important factor for evaluating concert hall acoustics.

IACC is measured using a small microphone installed at the entrance of the outer ear of a dummy head or real head. Two channel signals are recorded in the sound field of interest and then the signals processed by means of a digital computer. It must be noted that the measured value may change with the frequency spectrum of the source signal, so that standardisation of the method of measuring IACC will be required before its widescale application.

G. Total evaluation of a concert hall

Subjective studies of the acoustic characteristics of auditoria have shown that several parameters that can be obtained from measured impulse responses, correlated with particular subjective aspects of the acoustic character of an auditorium. While reverberation time gives the fundamental description of the acoustic character of an auditorium as a whole, the other room acoustic parameters give a more complete description of the acoustic conditions in the different seating areas of the auditorium.

Table 3.3 is an overview of the room acoustic parameters grouped into five different subjective listener aspects. The subjective level of the sound and the perceived reverberance are found to be particularly important for the evaluation of a concert hall. Although measurements are made in octave bands, it is appropriate to average the results in two or more octave bands as indicated in the table. When different results are to be compared, it may be useful to know the minimum step size that can be detected in listening tests, the Just Noticeable Difference (JND), which is also indicated in the table.

The typical range of the acoustic parameters is indicated for frequency-averaged values in non-occupied concert- and multi-purpose halls up to 25 000 m^3. In occupied halls, somewhat different values can be expected due to the additional absorption by the audience.

Table 3.3 Room acoustic parameters grouped according to their listener aspects

Subjective listener aspect	Acoustic quantity	Single number frequency averaging (Hz)	Just noticeable difference (JND)	Typical range
Subjective level of sound	Sound strength, G, in dB	500–1000	1 dB	−2 dB; +10 dB
Perceived reverberance	Early decay time, EDT, in s	500–1000	Rel. 5%	1.0 s; 3.0 s
Perceived clarity of sound	Clarity, C_{80}, in dB	500–1000	1 dB	−5 dB; +5 dB
	Definition, D	500–1000	0.05	0.3; 0.7
	Centre Time, T_S, in ms	500–1000	10 ms	60 ms; 260 ms
Apparent source width, ASW	Early lateral energy fraction, LF	125–1000	0.05	0.05; 0.35
Listener envelopment, LEV	Late lateral sound level, LG, in dB	125–1000	Not known	−14 dB; +1 dB

This extract is reproduced with the permission of the Danish Standards Foundation. The extract is based on the Danish Standard DS/EN ISO 3382-1:2009. © Danish Standards Foundation.

So far we have discussed how to evaluate room acoustics by the objective measurement of various physical quantities; however, subjective evaluation of the room acoustics as perceived by the human ear cannot be avoided.

For speech, a simple method of expressing comprehension by a percentage articulation test was described in Section 1.14. This is considered as an ideal evaluation method since a single numerical figure expresses the speech hearing condition, taking into account all influencing factors.

For music, on the other hand, the problem is so complicated that no definitive method has yet been established. But we can at least discuss a few proposed methods.

Beranek (1962) selected several evaluation factors for musical hearing. He proposed a method in which the attributes of acoustic quality were categorised and the physical quantities which might govern those selected attributes measured. Each was given a rating number on a rating scale as specified in Table 3.4 and the total score used to evaluate the hall. He collected the measurement data of 54 major concert halls from all over the world and concluded that the evaluations agreed well with the judgement of most musicians and critics (see Lit. A4). However, there are still many problems relating to the particular choice of attributes and the determination of rating scales. It was most unfortunate that the New York

Table 3.4 Rating scales for orchestral concerts (Beranek 1962)

Attribute	Physical quantity	Rating points	Max. point
Intimacy	Initial time delay gap (ms)	0–20 ms → 40 pts. → 70 ms → 0 pts.	40
Liveness	Reverberation time(s) at mid frequencies (500–1000 Hz for fully occupied hall)	Romantic 2·2 s, Typical orchestra 1·9 s, Classical 1·7 s, Baroque 1·5 s, (either longer or shorter time than the above give lower pts.)	15
Warmth	Average of RT at 125 and 250 Hz divided by RT at mid frequency	1·2 to 1·25 (larger or smaller than this range gives lower pts.)	15
Loudness of the direct sound	Distance from listener to conductor in ft	60 ft (beyond which every further 10 ft gives a reduction of 1 pt.)	10
Loudness of reverberant sound	$\dfrac{\text{RT at 500–1000 Hz}}{\text{room volumes ft}^3} \times 10^6$	3·0 (either larger or smaller gives lower pts.)	6
Diffusion	Wall and ceiling irregularities	If adequate	4
Balance and Blend	(Sectional balance in orchestra)	If good	6
Ensemble	Performers ability to hear each other	If easy	4

Philharmonic Hall, built in 1962 and acoustic designed on the basis of this method, was unsatisfactory.

PROBLEMS 3

1　Enumerate the various factors which play a role in the evaluation of the acoustic environment in a room and explain the evaluation methods.
2　Describe the principal features of geometrical and wave acoustics and explain how they are applied to problems in architectural acoustics.
3　Enumerate the phenomena to be avoided in room acoustics and indicate which is the most harmful.
4　When designing a class room whose floor area is about 55 m² with 3 m ceiling height, explain which is acoustic preferable for the floor plan, 6 m × 9 m or 7 m × 8 m.
5　Enumerate the factors to be taken into account in selecting the optimum reverberation time.

4 Sound absorption
Materials and construction

The absorption coefficient is a useful concept when using geometrical acoustic theory to evaluate the growth and decay of sound energy in a room. Any material absorbs sound to some extent. However, when sound is considered as a wave motion it is necessary to use the concept of acoustic impedance. In this chapter the fundamental characteristics of the terms which describe sound absorption, an outline of the performance of various absorbing materials, details of construction and their practical application in architecture are discussed.

4.1 Types of sound absorption mechanisms

Absorptive materials and construction can be divided into three fundamental types as shown in Figure 4.1.

A. *Porous absorption*

When sound waves impinge on a porous material containing capillaries or continuous airways, such as are found in glass wool, rock wool and porous foam, they propagate into the interstices where some of the sound energy is dissipated by frictional and viscous losses within the pores and by vibration of small fibres of the material. The absorption is large at high frequencies and small at low frequencies.

B. *Panel and membrane type absorption*

An impervious material, such as a thin plywood panel, painted canvas, etc., causes sound-induced vibration. A portion of the energy incident upon the material is dissipated by the internal loss of the vibrating system. The sound absorption characteristics produce a peak in the low frequency range; however, the peak is not very high unless a porous material is inserted in the back cavity.

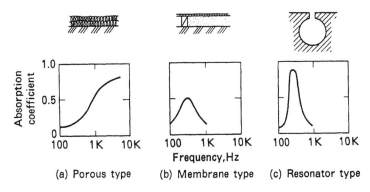

Figure 4.1 Absorption mechanisms and characteristics outline.

C. *Resonator absorption*

Sound incident on a resonator, consisting of a cavity with an opening, excites large-amplitude air vibrations at the opening in the resonant-frequency range, dissipating the sound energy by means of viscous losses. Thus, the absorption can be very large at the resonant frequency.

In practice these three types, or a combination of them, are considered effective as absorptive building materials or as part of a construction.

4.2 Measurement of absorption coefficient and acoustic impedance

The absorption coefficient depends on the type of sound incident on the material surface. As shown in Figure 4.2 when a plane wave is incident normally, on the material surface, i.e. when the angle of incidence is zero, the absorption coefficient is called the normal-incidence absorption coefficient and is denoted by α_0. In the case of oblique incidence, the coefficient is expressed as α_θ. For sounds incident at all angles, the coefficient is called the random-incidence absorption coefficient, while the calculated value of the statistical mean as a function of the angle of incidence is written α_s and the value obtained by measurement in a reverberation chamber is

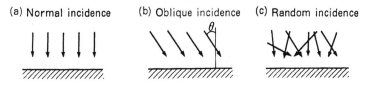

Figure 4.2 Sound incidence conditions.

called the reverberation (or reverberation room method) absorption coefficient α_r. The latter is generally designated α and used in practice for room acoustics and noise control. Although these absorption coefficients and acoustic impedances are defined at the surface of the material, it should be noted that they may also vary according to the thickness of the material and the backing condition.

A. Normal-incidence absorption coefficient and acoustic impedance

a. Measurement of normal-incidence absorption coefficient

Since a tube whose diameter is smaller than the wavelength is used to create a plane wave incident normal to the specimen, the method is called the tube method.

As shown in Figure 4.3, when the specimen is placed at the end of the tube and a pure tone is generated at the other, a standing wave is produced. The standing-wave pattern has peaks and troughs alternately spaced at distances of $\lambda/4$ according to Equation (1.35). Since the reflected sound amplitude B is small compared with the incident sound amplitude A, even at the position of pressure minimum in the reverse phase, $P_{min} = |A - B|$ never becomes zero. At the pressure maximum $P_{min} = |A + B|$, then

$$\frac{P_{max}}{P_{min}} = \frac{|A+B|}{|A-B|} = n$$

which is the standing-wave ratio. Moving a microphone (or probe-microphone in case of a slender tube) in the tube and measuring n, enables the sound pressure reflection coefficient $|r_p|$ to be determined.

$$|r_p| = \frac{|B|}{|A|} = \frac{n-1}{n+1} \tag{4.1}$$

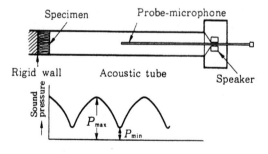

Figure 4.3 Standing-wave method.

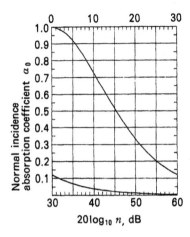

Figure 4.4 Relationship between 20 log n (dB) and α_0.

Therefore, the normal-incidence absorption coefficient is expressed as follows

$$\alpha_0 = 1 - |r_p|^2 = 1 - \left(\frac{n-1}{n+1}\right)^2 = \frac{4}{n+(1/n)+2} \tag{4.2}$$

This can be obtained from Figure 4.4 and also Table A.1 in the Appendices.

b. Measurement of acoustic impedance

When closely observing the pattern of a standing wave produced near the material surface, as shown in Figure 4.5, it can be seen that the distance from the material surface to the first pressure minimum P_{min} is

$$d = \frac{\lambda}{4} \pm \delta \tag{4.3}$$

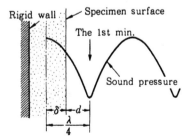

Figure 4.5 Standing-wave pattern.

The figure shows the case of $-\delta$. Thus, the position of the first $P_{\min}$, which occurs at $\lambda/4$ in the case of perfect reflection, is found to be displaced by δ. This is because the phase is shifted at the reflection point as the reflected wave is considered to start from the material surface; then the phase shift is

$$2k\delta, \quad (k = 2\pi/\lambda)$$

due to reflection.

Now, expressing the sound pressure reflection coefficient of the material surface in complex form with

$$\frac{B}{A} = |r_p| e^{j\Delta}$$

it follows that

$$\Delta = 2k\delta = 4\pi \frac{\delta}{\lambda} \tag{4.4}$$

Therefore, when n and δ are measured, $|r_p| e^{j\Delta}$ can be obtained by using Equations (4.1) and (4.4).

Using Z for the acoustic impedance of the material surface from Equation (1.30) it follows that

$$|r_p| e^{j\Delta} = \frac{Z - \rho c}{Z + \rho c} = \frac{z - 1}{z + 1} \tag{4.5}$$

where z is the impedance ratio. Hence,

$$\frac{Z}{\rho c} = z = \frac{1 + |r_p| e^{j\Delta}}{1 - |r_p| e^{j\Delta}}$$

Thus Z can be calculated. The relationship between $Z/\rho c$ and α_0 is also given by Equation (1.34). In practice the acoustic impedance ratio is expressed as follows

$$\frac{Z}{\rho c} = \frac{R}{\rho c} + j\frac{X}{\rho c} = r + jx \tag{4.6}$$

The real part r and imaginary part x can then be simply obtained by using a Smith chart (see Section 11.11) with measured values of n and δ shown in Figure 4.5.

Referring also to Equations (4.1) and (4.5), if $\Delta = 0$ then $n = z$. Therefore, when assuming that the reflecting surface is at the position shifted by δ from the material surface (see Figure 4.5), the phase shift of the reflected sound becomes nil, with the result that the impedance of the assumed reflecting surface is a real number n.

The diameter D of the measuring tube should satisfy the condition $D < 0.59\lambda$, otherwise the requirement for normal incidence for a plane wave will not be satisfied. Also, since P_{min} requires more than two measurements, the tube length needs to be larger than $3/4\,\lambda$. Therefore, two or three different sizes of tube are used to cover the frequency range of interest.

Recently the transfer function method has been widely used and measurement equipment according to this method is commercially available. In this method, a loudspeaker installed at one end of the impedance tube emits a wideband noise, and the transfer function between two microphones fixed in the tube is measured. The acoustic impedance of the specimen surface is calculated from the transfer function.

These tube methods have several advantages: small specimen size, simple apparatus for measurement and reproducible accuracy. Therefore, these methods are used for the comparison of materials and for developing new materials but they cannot be used for measuring absorption by panel/membrane absorbers.

B. Oblique incidence absorption coefficient

a. Measuring method

Although it is generally difficult, there are several methods for determining the oblique incidence absorption coefficient. The first is an analytical method using standing-wave modes in a rectangular room into which the specimen under investigation has been introduced (Hunt 1939). The second is a method based on steady-state measurements of the interference pattern caused by waves incident on and reflected from a specimen of sufficient size located in free space (Ando 1968). The third is a method of separating reflected and incident sound as follows:

(a) separation in space using the directivity of a measuring microphone,
(b) separation in the time domain with a short tone or an impulse response,
(c) separation based on an application of correlation techniques.

The new method of sound intensity measurement may also prove useful.

Figure 4.6 shows an example of the second method based on the interference pattern, where, at low frequencies, the accuracy is affected by the free-space dimension and the specimen size and at high frequencies by destructive interference. Also, when the angle of incidence gets close to 90°, the error increases.

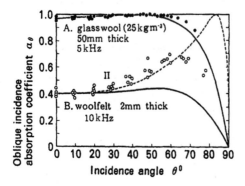

Figure 4.6 Oblique incidence absorption coefficients comparison of values measured by the interference pattern method with theoretical values, solid line: calculated by Equation (4.9), dotted line: values assuming $x_n = 0$ in Equation (4.9) (Ando 1966).

b. Assumption of locally reacting condition

At the boundary between two different media whose characteristic impedances are Z_1 and Z_2, the sound pressure reflection coefficient r_p can be determined against the angle of incidence θ_i as shown in Problem 1.2; therefore the absorption coefficient is expressed in a similar way to Equation (1.33) as follows

$$\alpha = 1 - \left| r_p \right|^2 = 1 - \left| \frac{Z_2 \cos\theta_i - Z_1 \cos\theta_t}{Z_2 \cos\theta_i + Z_1 \cos\theta_t} \right|^2 \tag{4.7}$$

However, for further simplification, the concept of a locally reacting surface is introduced, which assumes that the component of particle velocity perpendicular to the surface depends only on the pressure at the surface and not on the angle of incidence. Hence, the normal acoustic impedance Z_n of the material is defined as the ratio of the sound pressure to the particle velocity normal to the surface. Assuming that Z_n is independent of the angle of incidence, then the measured value obtained by the tube method, as described previously, can be used. Hence, substituting $\theta_t = 0$, and ρc (characteristic impedance of air) for Z_1 and Z_n (normal acoustic impedance of the material) for Z_2 and, replacing θ_i in Equation (4.7) with θ, the following is obtained

$$\alpha_\theta = 1 - \left| \frac{Z_n \cos\theta - \rho c}{Z_n \cos\theta + \rho c} \right|^2 \tag{4.8}$$

Then, substituting for Z_n using Equation (4.6)

$$\alpha_\theta = \frac{4\, r_n \cos\theta}{(r_n \cos\theta + 1)^2 + (x_n \cos\theta)^2} \tag{4.9}$$

Thus, the oblique incidence absorption coefficient varies with the angle of incidence θ. An example using the above equation is shown as the solid curves in Figure 4.6.

C. Random-incidence absorption coefficient

a. Statistical absorption coefficient

When a plane wave is uniformly incident from all directions, averaging the coefficients in terms of the angle of incidence using Figure 3.7, the following is obtained.

$$\alpha_s = \frac{\int_0^{\pi/2} \alpha_\theta \sin\theta \cos\theta \, d\theta}{\int_0^{\pi/2} \sin\theta \cos\theta \, d\theta} = \int_0^{\pi/2} \alpha_\theta \sin(2\theta) \, d\theta \tag{4.10}$$

This is called Paris's formula. Substituting Equation (4.9) in the above, the following is obtained

$$\alpha_s = \frac{8r}{r^2 + x^2}\left[1 - \frac{r}{r^2 + x^2}\log_e\left\{(r+1)^2 + x^2\right\} + \frac{r^2 - x^2}{x(r^2 + x^2)}\tan^{-1}\left(\frac{x}{r+1}\right)\right] \tag{4.11}$$

As there may be cases where the assumption of a constant value for Z_n, regardless of the angle of incidence, is not satisfied, the above expression is not always appropriate.

London's (1950) concept of non-directivity of the sound pressure at the point of incidence leads to a different statistical proposal as follows

$$\alpha_s^* = \frac{\int_0^{\pi/2} \alpha_\theta \sin\theta \, d\theta}{\int_0^{\pi/2} \sin\theta \, d\theta} = \int_0^{\pi/2} \alpha_\theta \sin\theta \, d\theta \tag{4.12}$$

Assuming there is less energy incident from directions near 90° in practical situations, the range of integration in Equation (4.12) is also defined as 0–78°. This was originally proposed for calculating random-incidence sound-reduction indices, but has also been used for calculating random-incidence sound absorption coefficients. The absorption coefficient obtained by this method is called the field-incidence-averaged sound absorption coefficient.

b. Measurement of reverberation absorption coefficient

In a completely diffuse sound field, the sound is incident on the wall surface from all directions. Suppose a reverberation room has a volume V (m^3) with total surface area S (m^2) and reverberation time T_0 seconds. When a specimen S_r (m^2) is placed in the room and a reverberation time T_m is obtained, Sabine's formula for the reverberation time (3.26) yields the following:

$$T_0 = \frac{55.3\,V}{cS\bar{\alpha}}, \quad T_m = \frac{55.3\,V}{c\,(S_r\alpha_r + (S - S_r)\bar{\alpha})}$$

where α_r is the reverberation absorption coefficient for the specimen and $\bar{\alpha}$ is the average absorption coefficient of the room. From the above two equations, the following is obtained:

$$\alpha_r = \frac{55.3\,V}{c\,S_r}\left[\frac{1}{T_m} - \frac{1}{T_0}\right] + \bar{\alpha} \tag{4.13}$$

Therefore, when the reverberation times T_m and T_0 are measured, the absorption coefficient α_r can be obtained; $\bar{\alpha}$ is so small that it may often be neglected.

For this measurement a fairly large area ($\sim 10\,\text{m}^2$) of material is required. Since the construction detail to be employed in the laboratory should be precisely that used in the field, the absorption coefficients to be used for the estimation of reverberation time and noise control purposes must be the actual measured values. It is also important to note that absorption characteristics due to panel vibration can be obtained only by this method.

The most important condition in this technique is that the sound field in the reverberation room is completely diffuse. However, when a sample is installed in the room, the condition becomes difficult to satisfy; therefore, even the same specimen will produce different results when measured in different laboratories as shown in Figure 4.7. Thus, in order to reduce the differences, ISO 354 provides the following instructions.

1   The reverberation room should have a minimum volume of 180 m^3, preferably more than 200 m^3.
2   The specimen should cover an area of about 10 m^2, placed on the floor so that the borders do not lie parallel to any wall.
3   Diffusion is to be provided, for instance by using many diffusing panels as shown in Figure 4.8, or rotating paddles.

Irregular room shapes without any parallel walls are preferable. Moreover, in order to make the reverberation time of the empty room as long as possible, the inside wall surface must have a hard smooth finish using painted concrete, polished terrazzo or mosaic tiles so that $\bar{\alpha}$ is between 1 and 2%.

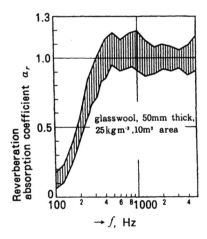

Figure 4.7 Measurement deviation range of absorption coefficients obtained by round-robin test in nine different reverberation rooms in Japan for the same specimen (1966).

c. Normal-incidence versus random-incidence sound absorption coefficients

The relationship between the normal-incidence absorption coefficient α_0 and the statistical absorption coefficient α_s depends upon how closely the assumption that the sample is locally reacting is satisfied. Furthermore, the relation of α_0 and α_s with the reverberation absorption coefficient α_r is affected by the diffusion characteristics provided in the reverberation room. In addition, the area effect described below has to be taken into account. As an example Figure 4.9 shows the relationship for measurements for dry sand and gravel.

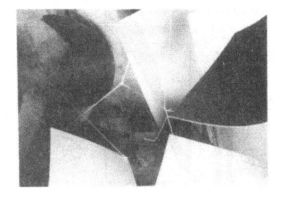

Figure 4.8 Diffusing panels hung in a reverberation room.

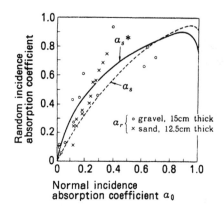

Figure 4.9 Relationship between measured values of α_0 and α_r and comparison with α_s (Maekawa 1962).

The curves are obtained from Equations (4.10) and (4.12) in which the following is substituted in place of Equation (4.8).

$$\alpha_\theta = 1 - \left(\frac{n\cos\theta - 1}{n\cos\theta + 1}\right)^2 \tag{4.14}$$

where n is the standing-wave ratio for normal incidence.

d. Area effect

When measuring the reverberation absorption coefficient, it is found that the smaller the specimen area, the larger the measured value, as shown in Figure 4.10. This is called the area effect and is more pronounced in cases where the linear dimensions of the specimen are less than the sound wavelength and also for materials whose coefficients are large. As a result α may exceed 1 (see Figure 4.7).

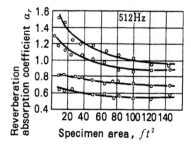

Figure 4.10 Area effect (Chrisler).

The area effect can be explained by the diffraction of the incident sound waves around the absorptive material. If only the absorbed part of the incident sound is considered, this can be compared to the case of an incident sound wave being transmitted through an opening with finite size. Especially in the case of a small aperture, it is a common observation that more sound is transmitted than would be expected from the area of the opening, and the angle of incidence has only a minor influence.

Considering the part of the incident sound that is reflected back from the surface, the relation between the sound pressure and the normal component of the particle velocity is given by the field impedance (or radiation impedance) of the surface, Z_f. In the ideal case of infinite area, the reflected sound is a plane wave, and the field impedance is a simple function of the angle of reflection θ.

$$Z_f = \frac{p}{v_n} \rightarrow \frac{p}{v \cos \theta} = \frac{\rho c}{\cos \theta} \quad \text{(plane wave)}$$

However, if a finite area is considered with the characteristic dimension e ($= 4 \times$area/perimeter), the field impedance can be approximated by the following, as found by Rindel (1993)

$$Z_f \approx \rho c \left[\left(\cos^2 \theta - 0.60 \frac{2\pi}{ke} \right)^2 + \pi \left(0.60 \frac{2\pi}{ke} \right)^2 + \left(\frac{2\pi}{(ke)^2} \right)^4 \right]^{-1/4} \quad (4.15)$$

The normalised field impedance calculated for a square area of $11 \, \text{m}^2$ is shown in Figure 4.11 as a function of the angle of incidence. At low frequencies it is found that the impedance is almost a constant ($= \rho c$), whereas at high frequencies the angle dependence gets closer to $\rho c / \cos \theta$ as for an infinite area, but only up to around 80° at 4000 Hz.

The relation between the absorption coefficient and the normal acoustic impedance of the material Z_n is then obtained by replacing Equation (4.8) with the following

$$\alpha_\theta = 1 - \left| \frac{Z_n - Z_f}{Z_n + Z_f} \right|^2 \quad (4.16)$$

Finally, the statistical absorption coefficient can be calculated using Paris's formula, Equation (4.10).

Using the equations above, the statistical absorption coefficient of a sound absorbing material with a small area, e.g. $1 \, \text{m}^2$, will get higher values than with the normal test area around $10 \, \text{m}^2$, especially at low frequencies. The calculated value may exceed 1.0, in agreement with the measured reverberation absorption coefficient. On the other hand, if the same material is installed in larger areas in a building, the area effect is reduced and may

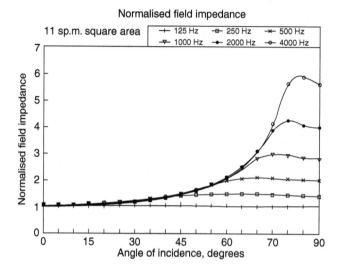

Figure 4.11 The field impedance normalised with ρc calculated from Equation (4.15) for an 11 m² square area. Reprinted from *Applied Acoustics*, 38, 2–4, J.H. Rindel, 'Modelling the angle-dependent pressure reflection factor,' 223–234, © 1993, with permission from Elsevier.

be neglected. For this reason it is recommended never to apply absorption coefficients higher than 1.0 in reverberation time calculations.

4.3   Characteristics of porous sound absorbers

A. Material thickness and air space

In a porous absorber, acoustic absorption is due mainly to viscous losses as air moves within the pores. Since the viscous loss is proportional to the dynamic pressure of the moving air, porous materials can provide more absorption when they are located in positions where the particle velocity of the sound wave is large. When sound is incident on a rigid wall, like concrete for example, a standing wave results and the particle velocity is a maximum at distances of $\lambda/4$, $3/4\,\lambda, \ldots$, from the wall as shown in Figure 1.9. On the other hand, Figure 4.12 shows the measured values of α_0 when cotton cloth is mounted with an air space in front of a rigid wall, and shows that even a thin porous material can provide considerable absorption where the particle velocity is large. In contrast, when a porous material is directly mounted on the wall where the particle velocity is zero, unless the material thickness is equivalent to $\lambda/4$, absorption is no longer effective. Within porous materials, the sound velocity is lower than in air and the wavelength becomes shorter. Therefore, generally the absorption

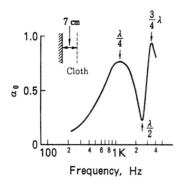

Figure 4.12 Measured α_0 for cotton cloth.

coefficient is low at low frequencies while at high frequencies the absorption coefficient is larger and is particularly large at those frequencies where the material thickness is equivalent to $\lambda/4$. From the above discussion, and as shown in Figure 4.13(a), it is clear that the thicker the material

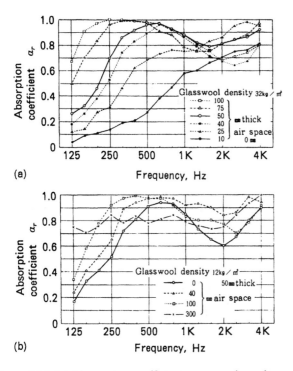

Figure 4.13 (a) Absorption coefficients measured on glass wool with various thicknesses. (b) Absorption coefficients measured on glass wool with various air spaces.

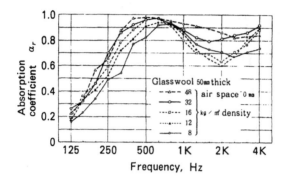

Figure 4.14 Absorption coefficients measured on glass wool with various densities.

the better the absorption at low frequencies and when improvement in absorption at low frequencies is required an air space should be provided as described in Figure 4.12. An example of the latter case is given in Figure 4.13(b). The material density is proportional to the price, but it suppresses the reduction of absorption over a wide frequency range as shown in Figure 4.14.

B. Material density and flow resistance

Regarding the density of porous materials, there is an optimum value for each species and frequency. For instance, the optimum density for rock wool is larger than that of glass fibre. This is mainly determined by flow resistance to air through the material. The flow resistance R_f is defined as

$$R_f = \frac{\Delta P}{u} \tag{4.17}$$

where the pressure difference ΔP (N m^{-2}) between both surfaces has been measured using the apparatus as shown in Figure 4.15 along with the flow speed, i.e. the particle velocity u (ms^{-1}) produced in the material.

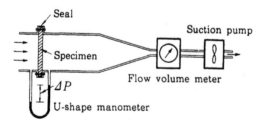

Figure 4.15 Measuring apparatus for flow resistance.

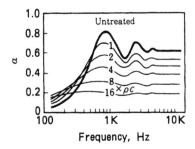

Figure 4.16 Relationship between flow resistance and absorption coefficient of porous material (Kosten 1960). Numbers on curves are the factors of the flow resistance of air in units of ρc.

R_f is given in N s m^{-3}, which is the same as Pa s m^{-1}, kg m^2 s^{-1} and the MKS rayl. Although this is analogous to direct current resistance, it is assumed applicable to a sound wave that is an alternating current and, hence, the term is often compared to the characteristic impedance of air ($\rho c = 415$ kg m^{-2} s^{-1}), which is often taken as the reference. An example is shown in Figure 4.16 in which, generally, the larger the flow resistance, the lower the absorption coefficient except at low frequencies where the situation is reversed.

The case shown in Figure 4.12 is confirmed by experiment, i.e. when the flow resistance equals ρc the theoretical value of the absorption coefficient is 100%.

C. Porosity and structure factor

Other factors governing absorptivity of porous materials are porosity and the structure factor. Porosity P is defined as follows.

$$P = \frac{\text{volume of pores connected to external air}}{\text{total volume}} \times 100\%$$

Foam-like materials, e.g. sponges, consist of different kinds of pores, i.e. connected and individual ones. While only connected pores are effective for absorption, individual ones barely absorb sound. Moreover, pore shapes and connection patterns are so varied that it may be appropriate to conceive a factor such as the structure factor to account for the deviation from the theoretical derivation based on the assumption of simple parallel capillaries.

Beranek (1947) proposed a method for evaluating the absorption by introducing a propagation constant (see Section 11.10, A), derived theoretically from an appropriate structure factor and measured flow resistance and porosity. This method is useful for research and the development of new materials (see Lit. B9a, B22).

Delany & Bazley (1970) proposed empirical formulas for predicting the propagation constant and specific acoustic impedance of a porous material from its flow resistance only. Their formulas are based on many experimental results and are valid for materials of high porosity. Miki (1990) proposed an improved version of Delany and Bazley's formulas, which are

$$z(f) = \frac{Z(f)}{\rho c} = 1 + 0.070 \left(\frac{f}{R_f}\right)^{-0.632} - j0.107 \left(\frac{f}{R_f}\right)^{-0.632}$$

$$\gamma(f) = k \left\{ 0.160 \left(\frac{f}{R_f}\right)^{-0.618} \right\} + jk \left\{ 1 + 0.109 \left(\frac{f}{R_f}\right)^{-0.618} \right\}$$

$$(4.18)$$

where $Z(f)$ is the specific acoustic impedance, $\gamma(f)$ is the propagation constant, f the frequency and $k(=2\pi f/c)$ the wave number.

D. Effect of construction on absorption characteristics of porous materials

Since glass fibre, rock wool, etc., are not, in themselves, suitable as surface finishes but require supplementary covering materials, the absorption characteristics may vary greatly depending on the nature of the latter.

1   Coarse and thin fabrics such as wire mesh, saran net, cheese cloth, etc. are transparent to sound, therefore they have no effect.
2   Perforated plates have little effect if thin, with many holes of small diameter and a perforation ratio more than 30%. When the perforation ratio is smaller, the absorption coefficient decreases at high frequencies. Then, depending upon the air space behind the plate, the absorber has characteristics more like those of a resonant absorber.
3   When an air-tight sheet or mat such as vinyl leather or canvas covers porous absorbents, the absorption at high frequencies decreases markedly for a hard porous material. However, the frequency curve shows distinct peaks and dips as shown in Figure 4.17 for soft porous material such as foam or cotton wool, with the additional effect that absorption is extended to lower frequencies.
4   Coating by oil paint blocks the pores and lowers the absorption, particularly at high frequencies. On the other hand, water paint applied so that no film is formed may be less detrimental.
5   In the case of board-like porous materials, such as acoustic tiles, if these are installed with sufficient air space behind then this allows the tile to vibrate and increases the absorption at low frequencies.

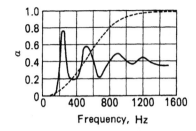

Figure 4.17 Absorption characteristics of mat type (Kosten 1960). 30-mm thick sponge rubber, dotted line: without any treatment; solid line: with air-tight sheet.

4.4 Characteristics of panel/membrane-type absorbers

An impermeable limp membrane or panel, such as a vinyl sheet, painted canvas, or thin plywood panel, which is placed parallel with a rigid back wall with an air cavity of depth L (m) in between, forms a single resonator composed of the mass of the membrane/panel and the stiffness of the air cavity. The resonance frequency is given as follows

$$f_r = \frac{1}{2\pi}\sqrt{\frac{\rho c^2}{mL}} = \frac{1}{2\pi}\sqrt{\frac{1.4 \times 10^5}{mL}}\,(\text{Hz}) \tag{4.19}$$

where m (kg m^{-2}) is the mass of the leaf per unit area, ρ is the air density, and c is the sound speed in air.

This equation gives the peak frequency of the absorption characteristics in the case of normal incidence of a plane sound wave.

Further, if the material possesses elasticity and permits the propagation of bending waves, this effect has to be added and the frequency expressed as follows, for a rectangular plate whose dimensions are $a \times b$ with a supported edge.

$$f_r = \frac{1}{2\pi}\sqrt{\frac{\rho c^2}{mL} + \frac{\pi^4}{m}\left[\left(\frac{p}{a}\right)^2 + \left(\frac{q}{b}\right)^2\right]^2 \frac{Eh^3}{12(1-\sigma^2)}}\,(\text{Hz}) \tag{4.20a}$$

where p, q are arbitrary positive integers, E is Young's modulus of the plate, h the thickness, and σ Poisson's ratio. If the plate is thin, i.e. h and m are small, the effect of the second term is reduced and, particularly, when L is small, the effect of the first term for the air spring dominates so that f_r from Equation (4.19) can be used as an approximate value. If, on the other hand, h and L become large, the resonant frequency is determined by the second term of the equation. The vibration mode examples are shown

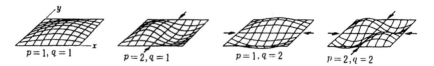

Figure 4.18 Normal mode vibration of plates.

in Figure 4.18 for various p and q values. Higher modes are not significant. The original equation was derived for idealised conditions, which may not be realised in practice; therefore, Equation (4.20a) can be expressed as follows:

$$f_r = \frac{1}{2\pi} \sqrt{\frac{1.4 \times 10^5}{mL} + \frac{K}{m}} (\text{Hz}) \tag{4.20b}$$

where K represents the panel's stiffness, and can be found experimentally. Examples of the values of K obtained for various types of panels are shown in Figure 4.19 and examples of measured absorption coefficients are illustrated in Figure 4.20.

The absorption peak in the oblique incidence case is generally observed at lower frequencies than Equation (4.20b). The peak frequency becomes lower as the air cavity becomes deeper, or as the panel/membrane becomes heavier. The absorption characteristics can therefore be adjusted by the air-cavity depth. However, when the mass of the panel/membrane or cavity depth is changed, not only the peak frequency but the peak value also changes. The absorption is caused by the internal loss of the vibrating system (including the panel's internal loss, supporting edges and sound absorption power inside the cavity) in absorbing some portion of the sound

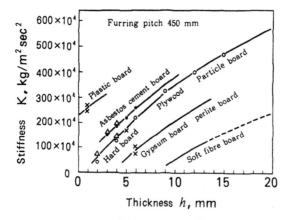

Figure 4.19 Stiffness of board panel (Sekiguchi *et al.* 1985).

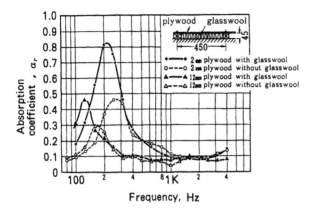

Figure 4.20 Absorption characteristics of panel vibration (Sekiguchi *et al.* 1985).

energy, but if there is no sound absorption power in the cavity, its absorption coefficient is not high.

The prediction of a sound absorption coefficient of this type is rather difficult, but recently some predicting theories have been proposed and discussed (Sakagami *et al.* 1996). Note that this type of sound absorption does not take place in the case of permeable leaves, which mainly show the porous-type absorption mechanism.

The following are some hints in the use of the panel/membrane-type of absorption:

1   Panels, which are thin and lightweight, provide more absorption, as they can vibrate easily. More absorption can be obtained if panels are nailed rather than bonded.
2   The absorption peak is larger if a porous material, such as fibreglass, is inserted in the back cavity. This also shifts the peak to lower frequencies.
3   Panels or membranes can be painted, as long as their vibration is not affected.
4   Panel/membrane-type absorption is basically caused when the panels are backed by a cavity and rigid wall because they absorb little sound energy themselves. However, double-leaf construction can also absorb sound energy to a certain extent, even if the back leaf is relatively lightweight (Kiyama *et al.* 1998).

There are some types of poroelastic hard panels that can cause both panel/membrane-type and porous-type absorption when placed with an air-back cavity.

4.5   Characteristics of single resonator absorber

Air in a cavity, whose dimensions are small compared with the wavelength, acts as a spring. When the cavity has a small opening to the outside air, the air in the neck moves as a single mass, the mechanical analogue of which is a mass supported by a spring, thus forming a simple resonator as shown in Figure 4.21. This is called a Helmholtz resonator. The resonance frequency is given by

$$f_0 = \frac{c}{2\pi} \sqrt{\frac{G}{V}} \tag{4.21}$$

where c is the sound speed, G is the air conductivity and describes the ease with which the air moves in the hole, and V is the cavity volume (see Section 11.2.D).

$$G = \frac{s}{l_e} \tag{4.22}$$

where s is the area of the opening and l_e is the effective neck length. In this context, where the air in the neck moves as a single mass, not only the air plug of length l but also portions of air at the front and back of the plug called the adding mass move, therefore, $l + \delta = l_e$ is used instead of the actual length l. δ is called the end correction. Then

$$f_0 = \frac{c}{2\pi} \sqrt{\frac{s}{V(l+\delta)}} \text{ (Hz)} \tag{4.23}$$

In the case of a circular hole whose diameter is d, $\delta \approx 0.8\,d$.

When sound at the resonant frequency strikes the resonator, the air in the neck vibrates strongly, thus absorbing the sound by viscous loss.

Such a single resonator is effective only for a limited range of frequencies close to the resonant frequency at which it shows a sharp peak. However, it is not suitable for use as a general absorber but is useful in controlling boom, resulting from a specific low frequency due to the natural mode of vibration of the air in the room (Figure 4.22). Although cavities can be

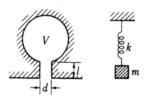

Figure 4.21 Single resonator.

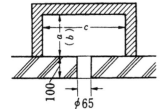

$a \times b \times c$	Resonant frequency
$150 \times 150 \times 300$	98 Hz
$200 \times 200 \times 400$	63 Hz

Figure 4.22 An example of applied single resonator (former Copenhagen Radio Hall ceiling, see Lit. B31).

made from timber, they are also made from tiles, bricks and ceramic jars or glass bottles and are built into the ceiling or wall, since the resonators should be both rigid and airtight. Using pipes for the neck enables f_0 to be controlled by the neck length. In order to increase the absorption, V should be increased but any dimension of the resonator must be less than about 1/6 of the wavelength concerned.

The resonator re-radiates sound that may not be heard because the decay rate in the resonator is greater than the room reverberation. Therefore, it functions only as an absorber.

4.6 Absorptive construction made of perforated or slotted panels

A. Resonant frequency due to perforated panel

In order to understand the behaviour of a perforated panel separated from a rigid wall by an air space, a system of air cells should be visualised with imaginary partitioning for the air space behind each hole of the panel (as shown in Figure 4.23), thus forming a series of Helmholtz resonators. The resonant frequency can be calculated from Equations (4.21) to (4.23). The hole opening ratio

$$P = \frac{\text{hole opening area summation}}{\text{whole panel area}}$$

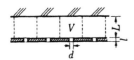

Figure 4.23 Absorptive construction with a perforated panel.

When the panel is L (m) from the wall and has n holes of open area s per unit area,

$$P = ns \qquad \therefore \qquad s = \frac{P}{n}$$

The cavity volume per unit area becomes L and the volume per hole is expressed as follows

$$V = \frac{L}{n}$$

Substituting these values into Equation (4.23), the following equation is obtained

$$f_0 = \frac{c}{2\pi} \sqrt{\frac{P}{L(l+\delta)}} \text{(Hz)} \qquad (4.24)$$

When the hole is circular, $\delta \approx 0.8\,d$ can be used.

B. Resonant frequency due to slotted panel

For a slotted panel, Equation (4.24) is applicable except that in this case the value of δ is given by

$$\delta = Kb \qquad (4.25a)$$

where b is the slot width.
    K is derived as follows.

1   *Limited slot length*: With length a

$$K = \frac{1}{\pi} + \frac{2}{\pi} \log_e \frac{2a}{b} \qquad (4.25b)$$

shown also in Figure 4.24 for values of a/b.
2   *Infinite slot lengths*: For a construction such as hit-and-miss boarding

$$K = \frac{2}{\pi} \log_e \left(\cos ec\frac{\pi}{2}P \right) \qquad (4.25c)$$

where the slot opening ratio $P = b/B$ and B is the slot pitch. K is obtained from Figure 4.25. Thus, the resonant frequency f_0 can be calculated from

$$f_0 = \frac{c}{2\pi} \sqrt{\frac{P}{L(l+Kb)}} \quad \text{(Hz)} \qquad (4.26)$$

for both examples of slot length.

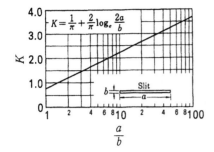

Figure 4.24 End correction K for limited slot length.

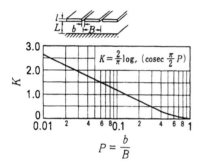

Figure 4.25 End correction K for infinite slot length.

C. Example of large air space

When the air space becomes larger and an array of single resonators is assumed, the cavities are considered as tubes instead of conceived as a simple spring; therefore, the closed pipe impedance should be employed. The frequency is then obtained as follows (Ingard & Bolt 1951), with the wavelength constant at the resonant frequency

$$\frac{l+\delta}{PL} = \frac{\cot(k_0 L)}{k_0 L} \qquad (4.27)$$

where

$$k_0 = \frac{\omega_0}{c} = \frac{2\pi}{c} f_0$$

After expansion of $\cot(k_0 L)^*$, only the first term is taken, Equation (4.27) becomes Equation (4.24). The difference is shown in Figure 4.26, which is convenient for the calculation of Equation (4.27).

$^*\cot x = \frac{1}{x} - \frac{x}{3} - \frac{x^3}{45} - \cdots$

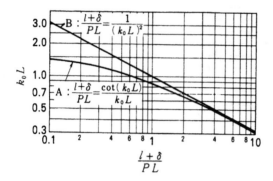

Figure 4.26 The resonant frequency for perforated or slotted panel construction.

[Ex. 4.1] When a panel of 8 mm thickness has very long slots of 5 mm width at 50 mm pitch, and the air space behind the panel is 30 mm thick, then from Figure 4.25 where $P = 5/50 = 0.1$, $K = 1.2$ is obtained. The resonant frequency is then obtained from Equation (4.26) as follows

$$f_0 = \frac{340}{2\pi} \sqrt{\frac{0.1}{0.03 \times (8 + 1.2 \times 5) \times 10^{-3}}} \approx 830\,\text{Hz}$$

When the air space is 300 mm thick, the left-hand side of Equation (4.27) gives

$$\frac{(8 + 1.2 \times 5)10^{-3}}{0.1 \times 0.3} = 0.467$$

From Figure 4.26, $k_0 L = 1.1$; therefore

$$f_0 = \frac{c \times 1.1}{2\pi L} = \frac{340 \times 1.1}{2\pi \times 0.3} \approx 200\,\text{Hz}$$

D. Absorption characteristics

(a) *Absorption characteristics due to normal incidence*: The absorptive system, which is composed of a perforated or slotted panel and air space, has an absorption coefficient which peaks at the resonance frequency obtained as in the calculation described above. The peak value and its spread are controlled by the resistance to air motion in the neck. Figure 4.27 shows the theoretical values for normal incidence and the relationship with the real part R of the impedance per unit area of the panel. When $R = \rho c$, $a_0 = 100\%$ at f_0. When $R < \rho c$, the peak value

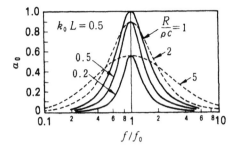

Figure 4.27 Resonant absorption characteristics due to perforated panel when the resistance at the neck is changed.

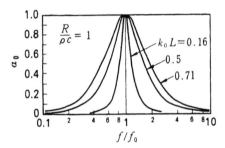

Figure 4.28 Resonant absorption characteristics due to perforated panel when the resistance at the thickness of air space is varied.

becomes lower but its spread is wider. R is proportional to the flow resistance in the neck and can be adjusted by cloth of appropriate thickness or by porous material such as glass wool behind the hole. When the resistance is optimum, the factor that controls the peak width is $k_0 L$, as shown in Figure 4.28; thus, the larger the air space, the wider the frequency range to be absorbed.

(b) *Absorption characteristic due to random incidence*: Generally, in the case of random incidence, by adjusting the dimensions of the hole opening and slit, thickness of air space and the flow resistance at the neck, there is the possibility of designing any arbitrary absorption characteristic (Ingard & Bolt 1951). The difference from the case of normal incidence is that when $R/\rho c$ is between 1.5 and 1.8 the absorption at f_0 becomes the maximum. As of now, although there are still problems in designing the resistance, fairly accurate prediction is possible by reference to actual measurements of various reverberation absorption coefficients.

As an example, when the air space is quite large, f_0 calculated from Equation (4.27) becomes a lower value and the statistical incidence absorption coefficients calculated from Equation (4.11) yield a series

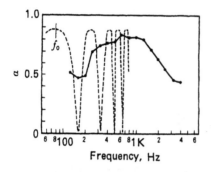

Figure 4.29 Absorption characteristics of perforated panel (hole opening ratio 13%) with large air space (1 m thick) (Sekiguchi *et al.* 1985).

of peaks lined up in the higher frequencies as shown by dotted lines in Figure 4.29, while the reverberation absorption coefficients, due to actual measurements, show a spectrum enveloping these peaks, thus producing a wide range absorption.

E. *Micro-perforated materials*

Recently micro-perforated panels (MPPs), which are perforated panels with submillimetre holes, are widely known as the most promising next-generation sound absorption material. MPPs and their theoretical basis were first proposed by Maa (1975). Their practical applications have since been studied by many researchers.

The absorption mechanism of MPPs is basically the same as ordinary perforated panels. Therefore, an MPP is used in the same way as ordinary perforated panels: placed parallel with a rigid back wall and an air-back cavity in between. However, with perforations smaller than a millimetre, MPPs can produce an acoustic resistance around ρc, which is optimal to obtain high absorptivity. Maa proposed the formulas predicting the acoustic resistance and reactance of an MPP as follows (Maa 1998)

$$r = \frac{32\eta}{\rho\rho c}\frac{t}{d^2}\left(\sqrt{1 + \frac{x^2}{32}} + \frac{\sqrt{2}}{32}x\frac{d}{t}\right)$$

$$\omega m = \frac{\omega t}{\rho c}\left(1 + \frac{1}{\sqrt{1 + \frac{x^2}{2}}} + 0.85\frac{d}{t}\right)$$

(4.28a)

where

$$x = \frac{d}{2}\sqrt{\frac{\rho\omega}{\eta}}$$

(4.28b)

and r is the acoustic resistance, ωm is the acoustic reactance (both normalised to ρc), d, t, and p are the hole diameter, panel thickness (throat length), and perforation ratio of an MPP, respectively. η is the coefficient of the viscosity of air ($\eta = 17.9\,\mu\text{Pa s}$).

Using these formulas, combined with the acoustic reactance of an air-back cavity, $-j\cot(kD)$ where k is the wave number and D the depth of the air-back cavity, the absorption coefficient of the absorber is given as

$$\alpha = \frac{4r}{(1+r)^2 + (\omega m - \cot(kD))^2} \tag{4.29}$$

Some attempts to make new types of sound absorber with MPPs have been made, such as a double-leaf MPP without back wall proposed as a space absorber. Also MPPs have been used for muffling devices of duct systems (Wu 1997) or as resistance components in sound insulation structures (Yairi et al. 2003).

4.7 Comments on commercial products, design and construction

In order to make absorption effective, one must understand fully the absorption mechanisms and characteristics so that appropriate designs can be developed to meet the requirements. At the same time, it is necessary to produce correct design and construction details and to have careful field supervision.

A. Selection of absorbent

Although, in practice, the most useful data is that provided by reverberation absorption coefficients, it varies depending not only on the material itself but also, critically, on the chosen construction detail, therefore, unless measured values relates directly to the construction detail, the data should not be trusted as a means of selecting and applying a particular commercial product. Some measured data are given in Table A.2 of the Appendices.

B. Use of area effect

Instead of using an absorbing material to cover the whole area, it is better to divide it into smaller areas, thus increasing the total absorption of the material as well as increasing the diffusion of reflected sound.

C. Selection of places for absorptive treatment

An absorbent is more effective as a means of increasing the total room absorption, when applied to places such as the corners or the perimeter of the ceiling where the sound pressure is high. This applies particularly to the use of single resonators, the performance of which is highly dependent on their position in a room.

D. Comments on installation

1 *Porous type*: It is advantageous to have as large an air space as possible behind the material. For instance, if an absorptive fibreboard is free to vibrations like a panel, this enhances absorption at low frequencies. As shown in Figure 4.30, the absorption of wood wool cement increases at middle frequencies due to penetrating interstices and, hence, it behaves like a perforated panel with an air space. On the other hand, such wood wool cement, when cast into concrete with its interstices filled, may lose absorption.

2 *Membrane vibration type*: It is important to install panels so that vibration can easily occur. In some cases resonant frequencies can be dispersed by varying the timber furring spacing and the number of nails.

3 *Perforated panel*: When this covers a porous absorbent, the hole opening ratio should be larger than 20% and, moreover, it is advisable to have the ratio as large as possible and the panel thickness as small as possible. In the case of the resonator type, the hole size, plate thickness and air space, etc., must all be manufactured correctly as calculated. It is important to note that the inner porous material must reach the hole to produce resistance.

E. Painting, paper and cloth covers

These finishes must be carefully applied in order not to destroy the absorption mechanism of the material itself or its composition. Although panel or membrane types of absorbers generally can be used without regard to the selection of paint, in the case of porous or resonator types they must obviously not be blocked by the paint or finishing materials. In the case of a

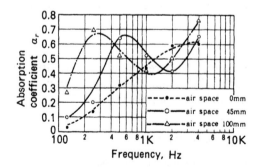

Figure 4.30 Absorption characteristics of cemented excelsior (wood wool) board (15 mm thick).

cloth covering, if the flow resistance is large, the material should be selected with care because it may cause a serious change in the absorption characteristics. The use of craft paper applied with paste should be avoided because air flow is greatly restricted.

Perforated panels often show dirt around the holes due to moving air depositing dust, particularly if cloth is used. This may cause serious aesthetic problems. The situation can be improved by placing a thin polyethylene film on the furring members behind the perforated panel. This reduces the deposition of dirt with little damage to the absorption because it gives little resistance to the alternating sound wave but blocks the direct flow of air.

4.8 Special sound absorptive devices

A. Variable absorbers

It is a natural desire to wish to change the reverberation time of a room to suit its purpose. Many devices have been conceived and developed over the years to provide variable absorption of walls and ceiling. However, it is difficult to maintain a complicated system and there are few technicians capable of managing such systems. Among others, draperies and movable panels, as shown in Figure 4.31 are increasing in popularity but still limited in application to special rooms such as broadcasting studios or listening rooms.

B. Suspended absorbers

When absorption is required for noise suppression in a factory or other situations, and where it is difficult to apply absorbents to walls and ceiling because of pipes, ducts and so on, it is quite effective to hang a number of such absorbers as shown in Figure 4.32 from the ceiling. Due to sound diffraction the apparent absorption coefficient may often become greater than 1. Another example of a suspended absorber is a system of numbers of discs made of perforated plate or wire mesh (whose diameter is 60–200 cm) with porous materials, and a deep grid composed of porous absorbent installed at ceiling height in auditoria, factories, etc.

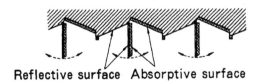

Reflective surface   Absorptive surface

Figure 4.31 Variable sound absorbing panels.

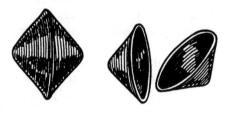

Figure 4.32 An example of a suspended absorber.

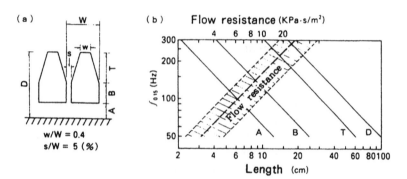

Figure 4.33 Design chart for economic absorptive wedge.

C. Absorptive wedge and anechoic room

A room in which the sound field is equivalent to open free space, is almost essential for any research into acoustics. This is achieved by complete absorption on all the surrounding surfaces of the room, which is then called an anechoic room. An absorptive wedge of porous material is used for this purpose.

Although the standard design of an absorptive wedge unit was presented by Beranek (1946), a new, more economic design is as shown in Figure 4.33 (Maekawa & Osaki 1983). Above the cut-off frequency indicated in the ordinate, the sound pressure reflection coefficients are lower than 0.15, provided that the flow resistance of the wedge material is controlled as shown in the figure.

If only frequencies higher than 300 Hz are of interest, wedges are not necessary; only a layer of glass wool, 20 cm thick, is required for lining the anechoic room.

PROBLEMS 4

1   Classify various sound absorptive materials and constructions in terms of their sound absorption mechanisms. Explain the general form of

their absorption characteristics and the caution necessary in their actual application.

2   In air, when a plane wave with pressure $p = Ae^{j(\omega t - kx)}$ is incident normally on a plane whose pressure reflection coefficient is $e^{-2(a+jb)}$, show that the standing wave which results can be described as follows

$$P(x) = 2Ae^{-(a+jb)} \cosh\left\{a + j(b - kx)\right\} e^{j\omega t}$$

and the standing-wave ratio is $n = \coth a$. Also show that the acoustic impedance density of the plane is

$$Z = \rho c \cdot \coth(a + jb)$$

for the above condition.

3   In a reverberation room whose volume is $150\,\mathrm{m}^3$ and surface area is $190\,\mathrm{m}^2$, the reverberation time at $500\,\mathrm{Hz}$ is $6.9\,\mathrm{s}$ when empty. When a specimen wall of $12\,\mathrm{m}^2$ area is installed in the room, the reverberation time is $3.4\,\mathrm{s}$. Find the absorption coefficient of the wall at $500\,\mathrm{Hz}$.

4   The reverberation absorption coefficient of the same absorbent produces different results when measured in different laboratories. Moreover, the results may exceed unity. Explain the factors that may account for this.

5   Glass wool, $50\,\mathrm{mm}$ thick, is mounted directly on a concrete surface. Compare the sound absorption characteristics in such an arrangement with those resulting from the introduction of a 100-mm air space behind the glass wool.

6   When 50-mm thick glass wool is mounted directly on a concrete surface, explain how the absorption characteristics will change with the following four surface finishes: (1) saran net; (2) vinyl leather; (3) 5-mm thick perforated panel with 4-mm diameter holes at centres 15 mm apart; (4) 5-mm thick perforated panel with 9-mm diameter holes at centres 15 mm apart.

5   Outdoor sound propagation

Noise control can be achieved by preventing the generation and spreading of noise. In this chapter, outdoor sound propagation and the effects of noise barriers are discussed.

5.1   Outdoor propagation of sound and noise

A. Attenuation due to distance

a. Attenuation due to distance from a point source

The sound intensity I at a distance d generated from a point sound source whose acoustic power W in a free field is given by

$$I = \frac{W}{4\pi d^2} \tag{5.1}$$

since the total energy that passes through a sphere of surface area $4\pi d^2$ is W.

Thus, I is inversely proportional to the square of the distance. The sound intensity/pressure level at this point can be expressed as follows

$$\begin{aligned} L &= L_W - 10\log_{10} 4\pi - 10\log_{10} d^2 \\ &= L_W - 11 - 20\log_{10} d \text{ (dB)} \end{aligned} \tag{5.2}$$

where the sound power level of the source L_W is $10\log_{10}(W/10^{-12})$.

If the sound source has a directivity factor Q (see Section 10.13) in the direction towards the receiving point

$$L = L_W - 11 - 20\log_{10} d + 10\log_{10} Q \text{ (dB)} \tag{5.3}$$

Since this equation shows that the inverse square law is also applicable in one direction, when the sound level at distance d_1 is L_1 (dB), L_2 at distance $d_2 = nd_1$, then it follows

$$L_2 = L_1 - 20\log_{10}\frac{d_2}{d_1} = L_1 - 20\log_{10} n \text{ (dB)} \tag{5.4}$$

Therefore, for every doubling of distance the sound level is reduced by 6 dB. Also L_W can be determined from Equations (5.2) and (5.3) with measured values of L, d and/or Q.

b. Attenuation due to distance from a line source

1   *Infinite line source*: When there are many sound sources in a row, such as vehicles on a motorway, they can be considered as an infinite line source. It is assumed that an infinite set of incoherent point sources lies continuously on the line in random phases, where any wave behaviour can be ignored. In a free field, the sound waves spread in a cylindrical form around a line source, which is the axis of the cylinder. When the acoustic power per unit length of the line source is W, the sound intensity I at distance d is given by Equation (5.5), since the sound energy is distributed over the cylinder surface with radius d.

$$I = \frac{W}{2\pi d} \tag{5.5}$$

In this case I is inversely proportional to d. When the source power level is L_W per unit length then the sound intensity/pressure level is given by

$$L = L_W - 8 - 10\log_{10} d \text{ (dB)} \tag{5.6}$$

Therefore, for every doubling of distance, the level is reduced by 3 dB.

2   *Finite line source*: In the case of a finite length line source, the point sources can be considered as lying continuously from x_1 to x_2 as shown in Figure 5.1. The energy density E at the receiving point P at a distance d from the source is

$$E = \int_{x_1}^{x_2} \frac{W dx}{4\pi r^2 c} = \frac{W}{4\pi c} \int_{x_1}^{x_2} \frac{dx}{(d^2 + x^2)}$$
$$= \frac{W}{4\pi c}\cdot\frac{1}{d}\left(\tan^{-1}\frac{x_2}{d} - \tan^{-1}\frac{x_1}{d}\right) = \frac{W}{4\pi c}\cdot\frac{\varphi}{d} \tag{5.7}$$

Thus, E is proportional to the angle φ, between the lines of sight from P to the respective sources x_1 and x_2 and inversely proportional to distance d.

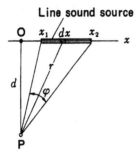

Figure 5.1 Line sound source.

[**Ex. 5.1**] In Figure 5.1, if the sound source length is l and $x_1 = x_2 = l/2$, the variation of sound energy density due to distance at P may be expressed with the aid of Equation (5.7) as follows:

$$E = \frac{W}{2\pi c} \cdot \frac{1}{d} \left(\tan^{-1} \frac{1}{2d} \right)$$

$$\left. \begin{array}{l} d \ll l; \quad \left(\tan^{-1} \dfrac{l}{2d} \right) \simeq \dfrac{\pi}{2} \quad \therefore E_{(near)} = \dfrac{W}{4c} \cdot \dfrac{1}{d} \\[3mm] d \gg l; \quad \left(\tan^{-1} \dfrac{l}{2d} \right) \simeq \dfrac{l}{2d} \quad \therefore E_{(far)} = \dfrac{Wl}{4\pi c} \cdot \dfrac{1}{d^2} \end{array} \right\} \tag{5.8}$$

If both values are made equal, $d = l/\pi$. We see from Figure 5.2 that if d is in the near distance compared to l/π, the sound pressure level reduces by 3 dB for every doubling of distance, while in the far distance it reduces by 6 dB for every doubling of distance. This approximation is very useful for estimating the sound distribution from a line source.

c. Attenuation due to distance from a plane source

When sound propagates through a window or wall, assuming an infinite set of point sources as were used to describe the line source, the energy density

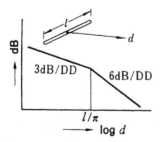

Figure 5.2 Distance attenuation from a line source.

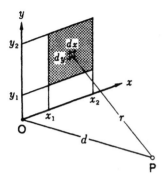

Figure 5.3 Rectangular plane source.

E at the receiving point P, distance d from a rectangular plane source as shown in Figure 5.3 is given by

$$E = \int_{x_1}^{x_2} \int_{y_1}^{y_2} \frac{W \cdot dxdy}{2\pi r^2 c} = \frac{W}{2\pi c} \int_{x_1}^{x_2} \int_{y_1}^{y_2} \frac{dxdy}{(d^2 + x^2 + y^2)} \qquad (5.9)$$

where W is the sound power per unit area.

Putting $x_1 = 0$, $y_1 = 0$, $x_2 = a$, $y_2 = b$ for simplification, the integral for unit distance d is derived as follows

$$\sum = \int_0^{a/d} \int_0^{b/d} \frac{dXdY}{1 + X^2 + Y^2} \qquad (5.10)$$

The result of the numerical integration is shown in Figure 5.4. Hence the sound intensity/pressure level at the receiving point P is

$$L = L_W - 8 + 10 \log_{10} \sum \text{ (dB)} \qquad (5.11)$$

Generally $\sum$ can be obtained as shown in Figure 5.5.

As in Figure 5.2, the distance attenuation can be approximated as shown in Figure 5.6, i.e. when d is less than a/π no attenuation occurs, while in the range $a/\pi < d < b/\pi$ attenuation may be approximated by 3 dB/DD, and in the range $d > b/\pi$ by 6 dB/DD.

B. Effect of reflecting surface

When a sound wave impinges on a smooth, hard, large surface there is a mirror-like reflection as shown in Figure 1.6 (Chapter 1).

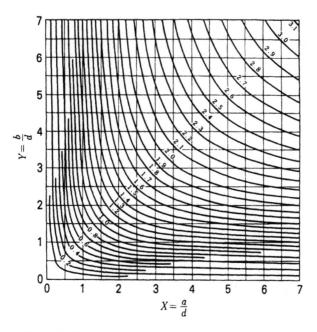

Figure 5.4 Calculation chart of Σ for a rectangular plane source.

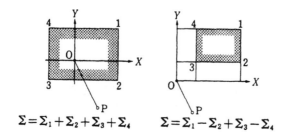

$$\Sigma = \Sigma_1 + \Sigma_2 + \Sigma_3 + \Sigma_4 \qquad \Sigma = \Sigma_1 - \Sigma_2 + \Sigma_3 - \Sigma_4$$

Figure 5.5 Method for obtaining Σ in general case.

a. Single reflecting surface

The reflected sound that is generated by the image source behind the reflecting surface, is added to the direct sound at the receiving point. In the case of pure tones their phase and, hence, the interference pattern, has to be taken into consideration, whereas, in the general case of noise, only the energy densities need to be considered. In Figure 5.7 the direct sound energy density at the receiver P is

$$E_0 = \frac{W}{4\pi r_0^2 c} \qquad (5.12)$$

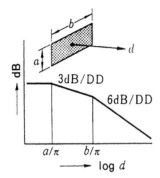

Figure 5.6 Distance attenuation from a plane source.

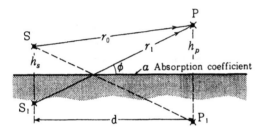

Figure 5.7 Sound reflection by a plane surface.

and the reflected sound energy density,

$$E_1 = \frac{W(1-\alpha)}{4\pi r_1^2 c} \tag{5.13}$$

added together it follows that

$$E = E_0 + E_1 = E_0 \left\{ 1 + \left(\frac{r_0}{r_1}\right)^2 (1-\alpha) \right\} \tag{5.14}$$

Alternatively, after obtaining each dB value for E_0 and E_1, the energy summation can be carried out using Figure 1.4. Another way would be to consider the reflected sound as propagating to the image P_1 of the receiving point from the sound source S, which produces the same result. This method is valid only when the reflecting surface is sufficiently large compared with the sound wavelength.

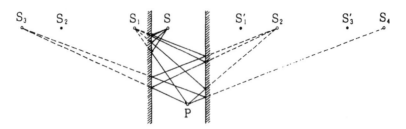

Figure 5.8 Multiple reflections between two parallel planes. Sound rays are shown from the 1st order image source at the left side, only.

b. Many reflecting surfaces

1   *The image sources.*

When there are many reflecting surfaces, each has not only an image source of first order but a second order image for a second order reflection, and also many higher order images corresponding to multiple reflections as shown in Figure 5.8. The energy density at the receiving point is obtained by integrating the energy densities from all images using the same principle as in Equation (5.14). Figure 5.8 shows an example of two parallel reflecting walls producing multiple reflections.

2   *Source close to a reflecting surface.*

   1   When a point source, whose power is known, is located close to one or more large reflecting planes, the following simplified method can be used, as shown in Figure 5.9.

      In the case of one reflecting surface the directivity factor is $Q = 2$, see Equation (5.3), because the energy spreads only within a hemisphere; therefore, the energy density is multiplied by 2.

   2   Similarly, when the sound source is located close to the orthogonal intersection of two reflecting surfaces, $Q = 4$.

   3   When the sound source is located close to the corner, where three reflecting surfaces intersect orthogonally, $Q = 8$.

[Ex. 5.2] When an air conditioning outlet is located on a ceiling, the difference in noise radiation between the centre and the corner locations can be

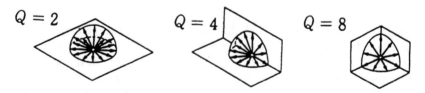

$Q = 2$ $Q = 4$ $Q = 8$

Figure 5.9 Directivity factors due to orthogonal intersection of reflecting planes.

found by using Equation (5.3). At the centre $Q = 2$ whereas at the corner $Q = 8$; therefore the difference is $10 \log (2/8) = -6 (dB)$, i.e. the noise level is reduced by 6 dB at the ceiling centre location.

C. Excess attenuation

Sound attenuation, with distance influenced by geometrical spreading and reflection as described above, is valid close to the source. At a distance of more than several tens of metres, the effects of the intervening medium meteorology and boundary conditions of the ground, etc., produce an excess attenuation greater than expected on the basis of the above formulas.

a. Attenuation due to air absorption

Sound waves attenuate during propagation in air because the energy is absorbed by the medium. When a plane wave, whose intensity is I, propagates a distance of x metres, the intensity I is given by

$$I_x = I_0 e^{-mx} \tag{5.15}$$

where m is the attenuation constant per metre. Expressing this in dB, the attenuation per metre is

$$10 \log_{10} (I_0 / I_1) = m 10 \log_{10} e = 4.34m \ (dB) \tag{5.16}$$

The value of m in air is given in Figure 3.10 (or ISO 9613-1).

Attenuation varies with humidity as well as temperature. However, at low frequencies, attenuation is so little that it can be neglected.

b. Influence of meteorological conditions

So far we have discussed the propagation of sound in stationary air. However, atmospheric air neither stands still nor is homogeneous. Meteorological conditions have a major influence on the propagation of sound in the open air.

Although fog or precipitation such as rain or snow have so little effect on sound propagation that they can be ignored, the profiles of air temperature and wind have a much greater influence.

1   *Effect of temperature profile*
    When the weather is fine during the day, the air is heated near the earth's surface by solar radiation but gets cooler towards the upper sky, called lapse. At night or in cloudy weather the temperature profile takes the form of an inversion. Since the sound velocity is greater when the temperature is higher according to Equation (1.7), the sound rays bend as

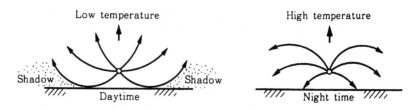

Figure 5.10 Sound refraction by temperature profile.

shown in Figure 5.10. Therefore, during the day, a shadow zone occurs beyond a certain distance from the sound source near the ground, where the sound is almost inaudible, while because of the higher temperatures in the upper air, the sound can easily return to earth at quite large distances from the source.

2   *Effect of wind*

Generally, wind speed is faster at high altitude than close to the earth. Since sound propagation is controlled by the vector summation of wind and sound speed, the sound rays bend as shown in Figure 5.11. Therefore, in the upwind direction shadow zones may occur while downwind sound reaches into the far distance. Figure 5.12 illustrates this phenomenon showing an abrupt change at midnight.

3   *Fluctuation of received sound level*

When sound propagates through the open air, there is a wide fluctuation in sound levels due to varying meteorological conditions with time at the receiving point, as shown in Figure 5.12. Furthermore, the natural wind blows so erratically with many irregular eddies that sound waves are forced to bend whenever striking eddies, thus leading to an absence of stable sound rays connecting source and receiver. Therefore, the received sound level shows violent variation from a very short duration to several minutes by 10 dB or more.

4   *Simulation of long-range sound propagation*

Sound propagation can be simulated by sound ray tracing with the aid of a computer, assuming steady meteorological conditions with constant profiles of temperature and wind between source and receiver. An example is shown in Figure 5.13 (see West *et al.* 1991).

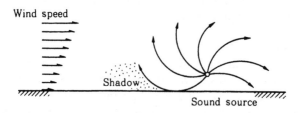

Figure 5.11 Sound refraction by wind speed profile.

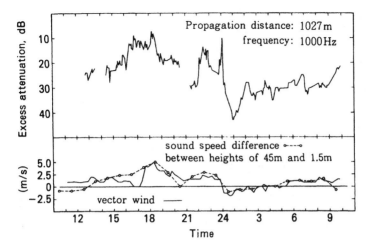

Figure 5.12 Measured fluctuation of receiving sound pressure level and vector wind (Konishi *et al.* 1979).

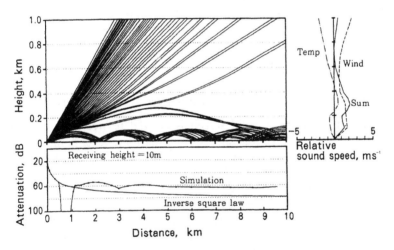

Figure 5.13 Sound ray simulation of long-range propagation (Takagi *et al.* 1985).

c. Attenuation due to ground effect

The excess attenuation of sound propagating near the ground is caused by the reflecting property of the ground. The reflection characteristic of the ground for a spherical sound wave is not defined by the reflection coefficient for a plane wave (see Section 1.5.C), because it is also requires a complicated mathematical function containing the distance from S to P, the grazing angle φ, and also the surface impedance Z, as shown in Figure 5.7.

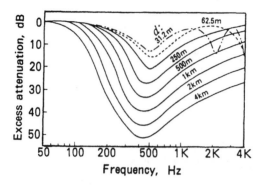

Figure 5.14 Excess attenuation due to ground covered by mown grass for $h_s = 1.8$ m, $h_p = 1.5$ m in Fig. 5.7. The values are relative to that for the point source placed on a perfectly hard surface (Piercy *et al.* 1977).

The curves in Figure 5.14 show excess attenuation for propagation over mown grass calculated by Piercy *et al.* (see Lit. B34). These results show good agreement with field measurements. The excess attenuation increases remarkably with distance at mid frequencies and depends on the surface impedance.

5.2 Noise reduction by barriers

A. Calculation of noise reduction by a barrier using a design chart

When a sufficiently large solid screen is erected between a noise source and receiver, although, on the source side, the noise level is raised due to reflection, on the receiving side a finite noise reduction can be expected because of the acoustic shadow. In order to obtain the sound pressure level in the shadow zone, the approximate diffraction theory of optics can be applied, though there are some discrepancies. Figure 5.15 shows the calculated and measured attenuations as well as experimental curves for a design chart (Maekawa 1968). This chart is derived on the basis of a semi-infinite screen in free space located between an omnidirectional sound source and a receiving point. The attenuation value indicates the difference between the situation with and without a screen. The abscissa in the figure shows the Fresnel number N, which divides the sound path difference with and without the screen into multiples of a half wavelength, and is adjusted so that the experimental values make a straight line. When $N = 0$, *SOP* is a straight line and when $N < 0$, the receiver is in the illuminated region; even so, there is a little attenuation, too. In the region $N > 1$ the attenuation can be expressed by $10\log(20N)$. This chart has been developed for easy design without use of a computer, as was the case in those days. If you prefer to use a computer, see Section 11.14.

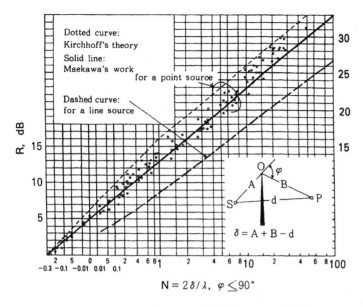

Figure 5.15 Sound attenuation by a semi-infinite screen in free space (Maekawa 1968).

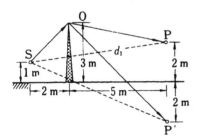

Figure 5.16 Example of noise reduction by a screen on the ground.

When a real screen is erected on the ground, the reflections from the ground and other objects must be considered. In practice, if the sound pressure level including the reflections is actually measured at the top O of the screen as reference, this approximates reasonably well to the sound level on the source side. On the receiving side, the calculation for the image P' of the receiving point P is also performed as shown in Figure 5.16 and their energies are summed, neglecting their phases as they are both noise signals.

[Ex. 5.3] When a 3-m high screen is erected between the sound source S and the receiving point P, as shown in Figure 5.16, the noise reduction can be obtained as follows: the path difference is

$$\delta_1 = \overline{SO} + \overline{OP} - d_1 = \sqrt{2^2 + 2^2} + \sqrt{1 + 5^2} - \sqrt{1 + 7^2} = 0.86\,\text{m}$$

The ground reflection is from path SOP'

$$\delta_2 = \sqrt{2^2 + 2^2} + \sqrt{5^2 + 5^2} - \sqrt{3^2 + 7^2} = 2.28\,\text{m}$$

Hence, N_1 for δ_1 is calculated, and the attenuation L is obtained from Figure 5.15. Similarly the attenuation L_2 is obtained from δ_2. Then, combining them, L_3 (dB) is obtained as shown in Table 5.1. This is the noise reduction of an omnidirectional sound source. At higher frequencies, when N is greater than one, it is clear that the attenuation increases by 3 dB oct^{-1}.

Table 5.1 Calculation of [Ex. 5.3]

f(Hz)	125	500	2000
λ (m)	2.72	0.68	0.17
$N_1 = \delta_1.2/\lambda$	0.63	2.52	10.10
$-L_1$ (dB)	−11.5	−17.0	−23.0
$N_2 = \delta_2.2/\lambda$	1.68	6.70	26.80
$-L_2$ (dB)	−15.0	−21.0	−27.0
$-L_3$ (dB)	−9.8	−15.5	−21.5

When the sound source is not omnidirectional and the sound pressure level measured at the top of the screen is L_0 then the sound pressure level at P is

$$L_p = \left(L_0 - 20\log_{10}\frac{d_1}{SO}\right) - L_3 \text{ (dB)} \tag{5.17}$$

Also, in the case of a line source such as a traffic stream on a motorway, the sound pressure level is approximated by

$$L'_p = \left(L_0 - 10\log_{10}\frac{d_1}{SO}\right) - L'_3 \text{ (dB)} \tag{5.18}$$

where L'_3 is the value calculated by the same method shown in Figure 5.16, except using the dashed curve in place of the solid line in Figure 5.15.

By these methods the sound pressure level at any point in the shadow zone of the screen can be calculated with good approximation, provided that

1 The sound pressure level distribution along the vertical line above the screen decays gradually away from the top O.
2 The wall length is several times larger than the wall height to either side of the receiver.

In order to fulfil condition (1), before erecting the screen, the vertical sound pressure level distribution should be measured upwards as high as possible and the position of O set higher than the maximum sound level position. In some particular cases, where the sound level increases as we progress upward, a noise reduction may not be expected. When condition (2) is not satisfied, diffraction, not only in the vertical direction but also laterally, has to be taken into account using Figure 5.15 (see Section 11.14).

B. Approximation for thick barriers

Until now it has been assumed that the screen has zero thickness. Barriers, however, like earth banks and buildings, have solidity. As a first approximation, in order to calculate the noise reduction, use the chart of Figure 5.15 with values of the path difference, as shown in Figure 5.17

$$\delta = \overline{SP} + \overline{OP} - \overline{SP}, \text{ or } \delta = \overline{SX} + \overline{XY} + \overline{YP} - \overline{SP}$$

This is the simplest way, though there are more accurate methods as described in Section 11.14.

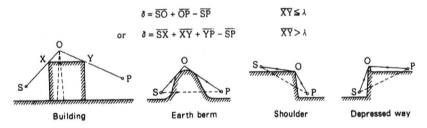

Figure 5.17 Approximation for thick barriers.

C. Treatment of large noise sources

It has been assumed that sound is propagated from a point source. If a large extended noise source is assumed to be an ensemble of many incoherent point sources, the sound energy received is obtained by summing the sound energies at the receiver from each point source as follows,

$$[\text{Att}] = 10 \log_{10} \left\{ \frac{\displaystyle\sum_{i=1}^{n} \frac{K_i}{d_i^2}}{\displaystyle\sum_{i=1}^{n} \frac{K_i}{d_i^2} \log_{10}^{-1} \frac{-[\text{Att}]_i}{10}} \right\} \ (\text{dB}) \tag{5.19}$$

where K_i is the power factor, d_i the distance to the receiver and $[\text{Att}]_i$ the value of attenuation due to the barrier for each point source (Maekawa *et al.* 1971).

For the special case of a street or a highway, the source can be treated as many point sources in a line parallel to the edge of the barrier. The result is shown by a dashed curve in Figure 5.15.

D. Effect of ground absorption

The calculation of sound attenuation by diffraction with a screen on the ground is performed in Figure 5.16, under conditions of perfect reflection at the ground surface. The ground, however, generally has a finite acoustic impedance, so that sound attenuates as it propagates over the ground without any barrier, as shown in Figure 5.14. If there is a barrier on the ground, the insertion loss of the barrier is obtained by subtraction of the ground attenuation from the barrier attenuation. The value of insertion loss may then, possibly, be negative. Figure 5.18 shows an example of the calculation of insertion loss by a barrier. We have to pay attention to the fact that the performance of a barrier on absorptive ground may be worse than expected (Isei *et al.* 1980).

E. Barrier effects of trees and forests

Although tree-plantings and forests furnish some attenuation due to absorption and dispersion, unless there is a substantial dense planting of considerable thickness (several tens of metres), there is very little attenuation. A quickset hedge, for example, has little to offer. However, tree-plantings are still absorptive compared with hard concrete or brick surfaces

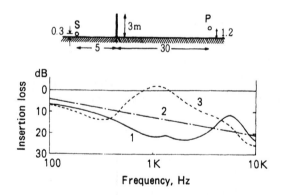

Figure 5.18 Calculated insertion loss of a barrier on the ground. The line source is above a hard surface, and the ground on the receiver side of the barrier is hard in Curves 1 and 2, but soft, grass-covered, in Curve 3. Curves 1 and 3 show exact results from wave theory, while Curve 2 is an approximate value estimated using the dashed curve in Figure 5.15 [Ex. 5.3].

and they do make both a visual and psychological contribution to the environment; the leaves, moreover, provide masking noise due to the wind; therefore, trees and shrubs can be considered as important elements in environmental noise control.

F. New trend of designing noise barriers

Noise reduction by screens in a human environment is not large unless it is on a huge scale. On the other hand, adverse effects due to other environmental factors, such as blocking sunshine, wind and sightlines, etc., may become too large to tolerate. Therefore, the effort should be made to reduce the adverse effects of the screen. In order to get greater noise reduction with lower height, many attempts to add or incorporate devices at the top of the barrier have been tried, (Okubo *et al.* 2010). Figure 5.19 (a) to (d) show an increasing number of diffraction edges, with (d) including some resonance. Further noise reduction can be increased by adequate sound absorptive treatment; (e) to (f) show increasing sound absorbing power, with an increasing amount of sound absorptive materials. (g) shows that the acoustic lens refracts noise from a direction incident from the right-hand side to the sound absorbing material. (h) shows the reactive surface on the top of a T-shaped barrier, i.e. an array of cylinders or wells, whose depth is equal to a quarter wavelength of the sound wave; the surface of the array is called 'acoustic soft', which means its impedance is zero for the frequency. Hence, theoretically, the sound pressure is zero, (h) for a single frequency and (i) to (j) aiming at a rather wide frequency range (Okubo & Fujiwara 1999). (k) shows a quadratic residue diffuser (QRD) (Figure 9.13) at the top of a T-shaped barrier. The performance of different surface

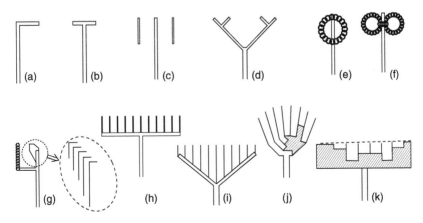

Figure 5.19 Special devices for the tops of noise screens to increase noise reduction.

treatments with and without absorbing materials has been researched only with numerical models (Monazzam & Lam 2008; Baulac *et al.* 2008).

Up to now, the devices have only been passive, but many authors are trying to apply active noise control (Chapter 8, Section 8.3.E.a.) for this purpose (e.g. Nakashima & Ise 2004).

As for sightlines, transparent materials are widely used. Today, almost all noise problems need noise barriers to deal with various kinds of transportation noise, such as highways, high speed railways and so on. These are large infrastructures in society. Thus, there should be an environmental impact assessment, not only for acoustics, but also in terms of the effect on the landscape. So, the aesthetics must be evaluated. There are now many attractive barrier designs in terms of plants, colour and architecture, which can be found throughout the world. (See Lit. A.23.)

PROBLEMS 5

1   By how much does noise attenuate from the centre of a window (W:10 m, H:2 m) of a factory, when the sound level is 93 dBA at the outer surface of the window? Calculate the sound level at six points, 0.5, 1, 2, 4, 8 and 16 m from the window using Equation (5.11) and compare with Figure 5.6.

2   There is a noise source of sound power level 100 dBA between two parallel, wide walls spaced 2 m apart, as shown in Figure 5.8. Calculate the sound level at a receiving point 3 m distance from the source, assuming a sound absorption coefficient of 0.8.

3   A noise screen is not always effective for noise reduction. What kind of noise is reduced effectively?

6 Airborne sound insulation

Airborne sound insulation is important for noise control in buildings, particularly when the noise source is speech, music or a noise source without mechanical connection to the building structure. In this chapter the fundamentals of sound transmission through walls, windows, openings and ducts are discussed.

6.1 Transmission loss

A. Sound transmission between rooms and between inside and outside

a. Sound transmission between adjoining rooms

The case where noise is transmitted from the source room to the receiving room through a partition, whose transmission coefficient is τ and area S is discussed. In Figure 6.1, when the energy density in the source room is E_1, the energy incident on the partition is $(c/4)E_1 S$ from Equation (3.20), so the energy transmitted into the adjoining room will be $(c/4)E_1 S\tau$.

If the energy density is E_2 in the receiving room, whose surface area is S_2, the energy incident on the whole surface is $(c/4)E_2 S_2$ and the absorbed energy $(c/4)E_2 S_2 \bar{\alpha}$ where $\bar{\alpha}$ is the average absorption coefficient. Since the total absorption $A_2 = S_2\bar{\alpha}$, in the steady state we obtain Equation (6.1)

$$\frac{c}{4}E_1 S\tau = \frac{c}{4}E_2 A_2 \quad \therefore \frac{E_1}{E_2} = \frac{1}{\tau}\frac{A_2}{S} \tag{6.1}$$

When the sound pressure levels in source and receiving rooms are L_1 and L_2, respectively, the level difference between the rooms is

$$L_1 - L_2 = 10\log_{10}\frac{E_1}{E_2} = 10\log_{10}\frac{1}{\tau} + 10\log_{10}\frac{A_2}{S}$$

$$\therefore L_1 - L_2 = R + 10\log_{10}\frac{A_2}{S} \text{ (dB)} \tag{6.2}$$

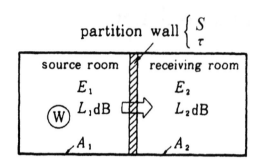

Figure 6.1 Sound transmission between adjoining rooms.

If the sound-source power is W in the source room with sound absorption A_1, expression (3.23) holds. From the definition of Equation (1.17) for L_W it follows that

$$L_1 = L_W + 6 - 10 \log_{10} A_1 \tag{6.3}$$

Then, substituting into Equation (6.2)

$$L_2 = L_W + 6 - R - 10 \log_{10} \frac{A_1 A_2}{S} \text{ (dB)} \tag{6.4}$$

Thus, not only the value of R of the partition wall, but also the sound absorption provided in both rooms is found to be important.

b. Sound transmission from outside to inside

Assuming that the incident sound on the outside wall is a plane wave whose intensity is I and the wall area is S, the incident energy is IS. In Figure 6.2, when the receiving room condition is the same as in Figure 6.1, with incident sound level L_0, the difference in sound pressure level is

$$L_0 - L_2 = R_0 - 6 + 10 \log_{10} \frac{A_2}{S} \text{(dB)} \tag{6.5}$$

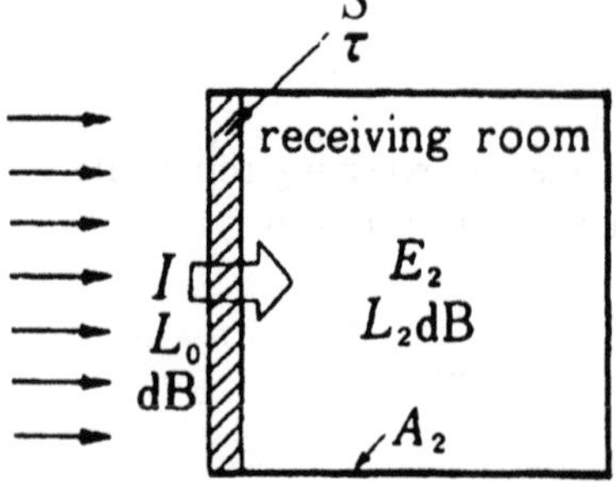

Figure 6.2 Sound transmission from outside to inside.

where L_0 must not include reflections from the wall. Also R_0 should be the value for normal incidence (see Section 6.3).

c. Sound transmission from inside to outside

When sound intensity I is radiated through the wall to the outside as shown in Figure 6.3, the radiated energy is IS. If the source room has the same condition as in Figure 6.1 and the sound level at the outside of the wall is L_0,

$$L_1 - L_0 = R + 6 \text{ (dB)} \tag{6.6}$$

substituting Equation (6.3) into Equation (6.6),

$$L_0 = L_W - 10\log_{10} A_1 - R \text{ (dB)} \tag{6.7}$$

where this L_0 value is at the outside surface of the wall and is equivalent to L_W in Equation (5.11). As the sound leaves the vicinity of the wall it spreads outwards.

B. Composite transmission loss

In actual buildings, walls are rarely constructed from a single material but contain windows and doors as well, thus, a wall often includes several components that have different R values. From Equation (1.25) the composite transmission loss denoted by R is expressed by,

$$\overline{R} = 10\log_{10}\frac{1}{\overline{\tau}} \tag{6.8}$$

where $\overline{\tau}$ is the average transmission coefficient, which can be obtained from the transmission coefficients τ_i for the area S_i

$$\overline{\tau} = \frac{\sum S_i \tau_i}{\sum S_i} = \frac{\sum S_i \tau_i}{S} \tag{6.9}$$

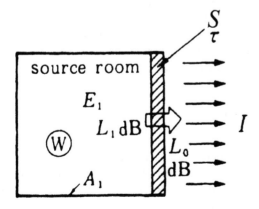

Figure 6.3 Sound transmission from inside to outside.

The transmission coefficient τ_i for the wall, whose transmission loss is R_i, is given by

$$R = 10 \log_{10} \frac{1}{\tau_i} \quad \therefore \quad \tau_i = 10^{-R_i/10} \tag{6.10}$$

Therefore, if each component transmission loss R_i is known, using the above three equations, the composite transmission loss can be calculated.

[Ex. 6.1] An exterior wall constructed from reinforced concrete has an area $30\,m^2$ in which there is a glass window of $10\,m^2$ and the R values are 50 and 20 dB, respectively. Then the composite transmission loss of the wall is calculated as follows

$$\text{Wall:} \tau_1 = 10^{-(50/10)} = 0 \cdot 00001, \quad \text{Window:} \tau_2 = 10^{-(20/10)} = 0 \cdot 01$$

$$\overline{\tau} = \frac{(30-10) \times 0 \cdot 00001 + 10 \times 0 \cdot 01}{30} = \frac{0 \cdot 1002}{30} \approx \frac{1}{300}$$

$$\overline{R} = 10 \log_{10} 300 = 24 \cdot 8\,\text{dB}$$

[Ex. 6.2] When $1\,m^2$ of the window is open in the above wall, the composite transmission loss will be

$$\overline{R} = 10 \log \frac{30}{20 \times 0 \cdot 00001 + 9 + 0 \cdot 01 + 1 \times 1} \approx 10 \log \frac{30}{1 \cdot 1} = 14 \cdot 4\,\text{dB}$$

6.2   Measurement and rating of airborne sound insulation

A. Measurement of sound transmission loss: reduction index

The test specimen is inserted in an opening between two adjacent reverberation rooms as shown in Figure 6.4. The sound is generated in one room, and the sound pressure levels L_1 and L_2 in both rooms are measured in the steady state. Then the sound transmission loss R is obtained from Equation (6.11) derived from Equation (6.2)

$$R = L_1 - L_2 - 10 \log_{10} \frac{A_2}{S}\,(\text{dB}) \tag{6.11}$$

where S is the area of the specimen, and A_2 is the absorbing area in the receiving room. For this purpose, sufficient structural isolation between source and receiving room is essential in order to be able to neglect the sound transmitted via any indirect path.

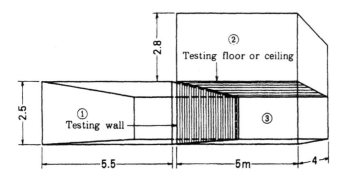

Figure 6.4 Reverberation rooms for measuring sound insulation, minimum dimensions recommended by ISO 140. (1) Source room for test wall (≈ 10 m²). (2) Source room for test floor and ceiling (10–20 m²). (3) Receiving room (>50 m³).

ISO 140 provides the following instructions:

1   The volumes and shapes of the two test rooms should not be exactly the same. The volumes should be at least 50 m³ and there should be at least 10% difference between them, in order not to have the same natural frequencies in both rooms. The room dimensions should be chosen so that the natural frequency should not degenerate in the low-frequency range as shown in Figure 3.5. Further diffusion is to be provided, if possible, by using diffusing elements as shown in Figure 4.8.

2   The reverberation time should be modified by not more than 2 s, especially at low frequencies, in order for the measured value not to depend on the reverberation time.

3   The test opening should be approximately 10 m² for walls and 10–20 m² for floors, with the shorter edge length not less than 2.3 m. The specimen should be installed in a manner as similar as possible to the actual construction.

4   Sound transmission via flanking paths should be negligibly small.

5   Sound pressure levels in both rooms should be measured at many points; then the average level L is obtained with the aid of Equation (1.22). White noise is used and the sound pressure level measured in 1/3 octave bands. The frequency range should be at least from 100 to 3150 Hz, preferably 4000 Hz.

6   The sound absorption area A_2 is obtained from the measured reverberation time using Equation (3.26).

Figure 6.4 shows the minimum size of measuring rooms according to the above specification.

B. Measured data of sound transmission loss

There have been many measured data already published for various materials and construction details. See Appendices Table A.3. These values are used in practical calculations of noise reduction. If you cannot get measured data for a particular material then you can estimate the R value by choosing the R value of a similar material in Table A.3, taking note of the comments in this chapter.

C. Field measurement of airborne sound insulation

When measuring airborne sound insulation in an actual building, there are flanking sound transmission paths as well as the direct path through the partition, as shown in Figure 6.5. Therefore, even if the measurement is carried out in the same way as in the laboratory, the value should be called the apparent sound reduction index, R'.

$$R' = L_1 - L_2 - 10 \log_{10} (A_2/S) \text{ (dB)} \tag{6.12}$$

This value may be used for comparison with the laboratory measured value of R.

In order to evaluate the airborne sound insulation between rooms in the field, the level difference

$$D = L_1 - L_2 \text{ (dB)} \tag{6.13}$$

is used.

However, the receiving sound pressure level is inversely proportional to the sound absorption area in the receiving room. So, the normalised level difference, with reference absorption 10 m^2, is given by

$$D_{n,10} = L_1 - L_2 - 10 \log_{10} (A_2/10) \text{ (dB)} \tag{6.14}$$

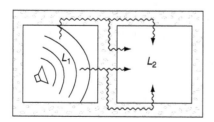

Figure 6.5 Sound transmission paths through walls between adjoining rooms.

Alternatively, normalising the measured reverberation time T_2 in the receiving room to a reference value of 0.5 s, which is typical of domestic rooms,

$$D_{n,0.5} = L_1 - L_2 - 10\log_{10}(T_2/0\cdot5)\ \text{(dB)} \qquad (6.15)$$

Equation (6.15) is adopted and called the standardised level difference by the International Standards Organisation.

D. Single number rating of airborne sound insulation

Generally, the value of R depends on frequency. However, it is often desirable to convert the information into a single number rating of acoustic performance. ISO 717-1 provides a method as follows:

1   The measured curve of R (or R′) values is compared with a reference curve shown in Figure 6.6.
2   The reference curve is shifted towards the measured curve until the mean unfavourable deviation is as large as possible but not more than 2.0 dB.
3   The value of the reference curve at 500 Hz, R_w (or R'_w) is called the weighted sound reduction index (or weighted apparent sound reduction index).
4   The same procedure is also applied to the value of D and $D_{n,0.5}$, then D_w and $D_{n,0.5w}$ are called the weighted level difference and the weighted standardised level difference, respectively.

Unfortunately, many countries have their own standards or codes with different criteria, although most systems are similar in principle. Some of them are shown in Section 11.12.

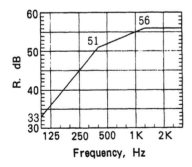

Figure 6.6 Curve of reference values for airborne sound (ISO 717-1).

6.3   Mass law for sound insulation of a single wall

A. Normal incidence mass law

When a plane wave, whose angular frequency is $\omega = 2\pi f$, is incident normally on an infinitely wide thin wall, some of the wave is reflected and some transmitted. Let the sound pressures of the incident, reflected and transmitted sounds be denoted by p_i, p_r, and p_t, respectively, as shown in Figure 6.7(a). The wall is excited by the sound pressure difference between the two surfaces of the wall and the equation of motion is

$$(p_i + p_r) - p_t = m\frac{dv}{dt} \tag{6.16}$$

where m is the surface mass of the wall, and v is its velocity.

In the case of simple harmonic motion, $d/dt = j\omega$ can be used (see Problem 1.1),

$$(p_i + p_r) - p_t = P = j\omega mv \tag{6.17}$$

$$\therefore j\omega m = \frac{P}{v} \tag{6.18}$$

This is the impedance per unit area of the wall. Since it is assumed that the particle velocity of the air adjacent to both wall surfaces is equal to v,

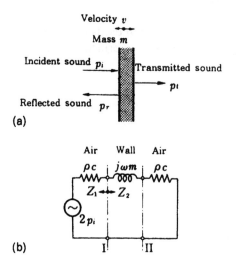

(a)

(b)

Figure 6.7 (a) Sound insulation of single wall and (b) its analogous circuit.

$$\frac{p_i}{\rho c} - \frac{p_r}{\rho c} = \frac{p_t}{\rho c} = v$$

$$\therefore\ p_i - p_r = p_t = \rho c v \tag{6.19}$$

From Equations (6.17) and (6.19)

$$\frac{p_i}{p_t} = 1 + \frac{j\omega m}{2\rho c} \tag{6.20}$$

Therefore the transmission loss is

$$R_0 = 10\log_{10}\frac{1}{\tau} = 10\log_{10}\left|\frac{p_i}{p_t}\right|^2 = 10\log_{10}\left\{1 + \left(\frac{\omega m}{2\rho c}\right)^2\right\} \tag{6.21}$$

Generally, $(\omega m)^2 >> (2\,\rho c)^2$; hence

$$R_0 \approx 10\log_{10}\left(\frac{\omega m}{2\rho c}\right)^2 = 20\log_{10} f \cdot m - 43\,(\text{dB}) \tag{6.22}$$

which is proportional both to frequency and the surface mass m of the wall. This is called the mass law for airborne sound insulation. Doubling the weight of the wall or the frequency gives an increase of 6 dB in R_0.

This wall motion may be expressed by an analogous electrical circuit as shown in Figure 6.7(b) (see Section 11.2). In this circuit, if no wall exists, it becomes as shown in Figure 6.8(a) and at the wall surface the sound pressure is p_i. When the wall is completely rigid, in which case the velocity is zero, the current is zero, thus the circuit is open as shown in Figure 6.8(b), and at the wall surfaces

$$p_i + p_r = p_i \ \ \therefore\ p_r = p_i$$

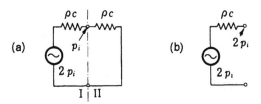

Figure 6.8 Analogous circuit: (a) open air, and (b) rigid wall.

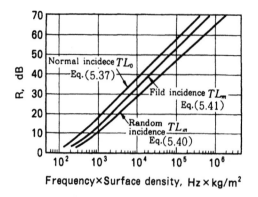

Figure 6.9 Mass law curves for transmission loss (Beranek 1954, Lit. B13).

In Figure 6.7(b), at the junction I of the circuit, the impedances to the left and the right are denoted by Z_1 and Z_2, respectively. Since there is no internal absorption, the transmission coefficient τ can be expressed by Equation (1.32), where $Z_1 = \rho c$, $Z_2 = j\omega m + \rho c$, then Equation (6.21) is obtained.

B. *Random incidence mass law*

When the incident angle is θ, Equation (6.23) can be derived as follows

$$R_\theta = 10\log_{10}\left(\frac{1}{\tau_\theta}\right)^2 = 10\log_{10}\left\{1 + \left(\frac{\omega m\cos\theta}{2\rho c}\right)^2\right\} \tag{6.23}$$

If we are calculating the average value in the range $\theta = 0\text{--}90°$, Equation (6.24) is obtained as the random incidence mass law

$$R_{\text{random}} = R_0 - 10\log_{10}(0{\cdot}23R_0) \tag{6.24}$$

However, in an actual sound field, using the range of $\theta = 0\text{--}78°$, is more realistic and the following approximate formula is obtained

$$R_{\text{field}} = R_0 - 5\,\text{dB} \tag{6.25}$$

which is recognised as closer to reality and called the field incidence mass law. Figure 6.9 shows these theoretical curves (see Lit. B16, B22).

6.4   Coincidence effect on sound transmission

The mass law has been derived on the assumption that the walls are set in uniform piston motion. However, flat plates are accompanied by bending vibrations, which cause significant decreases in R values.

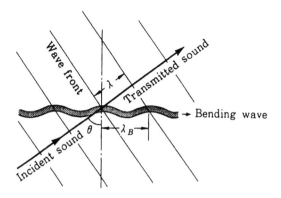

Figure 6.10 Coincidence effect.

As shown in Figure 6.10, when a plane wave whose wavelength λ is incident on a wall at angle θ, a pattern of alternating sound pressure moves along the wall with a wavelength of

$$\lambda_B = \frac{\lambda}{\sin\theta} \tag{6.26}$$

Therefore, the wall is excited by a bending vibration, which produces a bending wave that propagates along the wall surface.

On the other hand, the propagation speed c_B of the bending wave of a plate, whose thickness is h, can be derived from the theory of bending vibration of a bar.

$$c_B = \left(2\pi hf\sqrt{\frac{E}{12\rho(1-\sigma^2)}}\right)^{1/2} \tag{6.27}$$

where ρ is the density of the plate, E is the Young's modulus of the plate material and σ Poisson's ratio. c_B an be seen to increase with frequency (see Lit. B6, B19). In Figure 6.11 we can find the condition where the value of c_B satisfies Equation (6.28),

$$c_B = \frac{c}{\sin\theta} \tag{6.28}$$

At this frequency, the bending vibration is such that its amplitude may become comparable to the incident sound wave, resulting in a serious decrease in sound insulation. This frequency is called the coincidence frequency. This phenomenon is called the coincidence effect, as distinct from

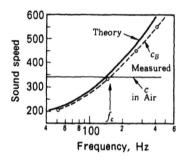

Figure 6.11 Propagation speed of bending wave (see Lit. B16).

resonance. The **coincidence frequency** is obtained from Equations (6.27) and (6.28) as follows

$$f_{(\theta)} = \frac{c^2}{2\pi h \sin^2 \theta} \sqrt{\frac{12\rho(1-\sigma^2)}{E}} \tag{6.29}$$

The lowest coincidence frequency when $\theta = 90°$ is given by

$$f_c \approx \frac{c^2}{2\pi h} \sqrt{\frac{12\rho}{E}} \approx \frac{c^2}{1 \cdot 8 h c_s} \tag{6.30}$$

where $\sigma = 0.3$ (leading to an approximation $(1-\sigma^2) \approx 1$) and c_s is the sound speed in the solid material as shown in Equation (1.8).

f_c is called the **critical frequency**. Any lower frequency than this yields $c_B < c$ resulting in an absence of coincidence, while at any higher frequency than f_c, coincidence will occur.

In Figure 6.12, the relationship between the material thickness h and f_c for various materials is shown. The higher f_c with smaller h the less the effect, but with larger h, f_c decreases to the middle or lower frequency range essential for good sound insulation and, so, the transmission loss is seriously decreased. Figure 6.13 shows an example where f_c is different depending upon the wall material even where the walls have the same surface density.

The decrease of R by coincidence is very complicated, as it is also related to the loss factor due to the material's internal friction.

Figure 6.14 is a practical design chart from a theoretical and experimental study, which gives good approximation for large panels. The length and width of the panel should be at least 20 times the panel thickness (Watters 1959).

The straight line portion (1)–(2) is drawn from the field incidence mass law Equation (6.25). The plateau height and breadth (2)–(3) are determined from the table. The part above (3) is an extrapolation (see Lit. B22).

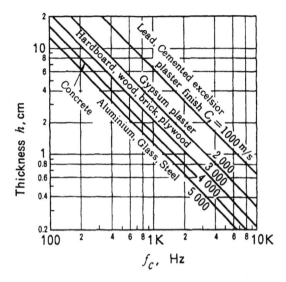

Figure 6.12 Critical frequencies vs material thickness.

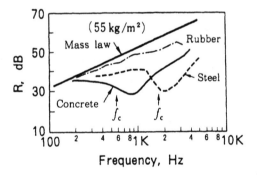

Figure 6.13 Coincidence effect of different materials with the same surface density (Cremer et al. Lit. B19).

6.5   Frequency characteristics of a single wall

We have considered the infinitely wide wall for the mass law and the effect of coincidence on the sound transmission. This may be a reasonable approximation for middle and high frequencies on a wall of finite size. In the low-frequency range, however, a finite wall has its own resonances, as shown by Equation (4.20a) and Figure 4.18 under the condition of a supported edge. Hence, the frequency characteristics of the airborne sound transmission loss might be as shown in Figure 6.15, although, if the real wall were rigidly clamped at the boundaries, the lowest resonance frequency f_{rl},

	Surface density $\mathrm{kg\,m^{-2}\,cm^{-1}}$	Plateau height dB	Plateau breadth f_3/f_2
Aluminium	26.6	29	11
Concrete, dense	22.8	38	4.5
Glass	24.7	27	10
Lead	112	56	4
Plaster, sand	17.1	30	8
Plywood, fir	5.7	19	6.5
Steel	76	40	11
Brick	21	37	4.5
Cinder block	11.4	30	6.5

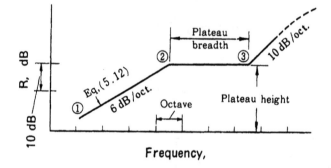

Figure 6.14 Practical estimation chart for R of a single panel taking into account coincidence (Watters 1959).

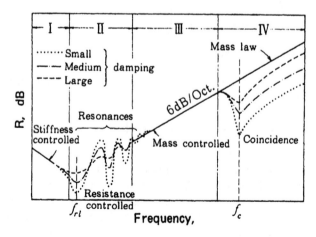

Figure 6.15 Frequency characteristics of the airborne sound insulation of a single wall (Maekawa 1990).

would be twice as high, and all others would be higher than those obtained from Equation (4.20a).

In region I in Figure 6.15, below f_{rl}, the wall is generally controlled by its stiffness and edge condition. In the second region, higher resonances are controlled, not only by the mass but also by internal energy losses and losses at the boundaries. At frequencies two to three times f_{rl} up to and near the critical frequency f_c, is the mass controlled region III. In region IV, there is multi-coincidence, which is controlled by both stiffness and resistance. It is very difficult to estimate the sound transmission in the lowest frequency region I, but easier to make approximations in the higher frequency regions as mentioned above (Figure 6.14).

6.6 Sound insulation of double-leaf walls

A. Construction of double-leaf walls

According to the mass law, the transmission loss of a single wall increases by only 6 dB for a doubling of the wall thickness, i.e. twice the mass (Figure 6.9). Also, when the thickness increases, the coincidence effect may cause undesirable results. This suggests that sound insulation of a single wall has its limitations. However, if the wall is made up of two leaves, each of which is an isolated single wall, the total transmission loss must be the sum of the transmission loss of each leaf.

For example, if R is 48 dB for a 15-cm reinforced concrete wall, the double-leaf wall with sufficient air space can give an R as much as 96 dB; on the other hand, if the wall thickness is doubled, i.e. to 30 cm, only 54 dB is obtained even if there is no coincidence effect. Therefore, a double leaf construction has considerable advantage over the single wall.

In practice, it is very difficult to make the two leaves completely detached, so that sound is transmitted via (1) the structural coupling and (2) the acoustic coupling due to the air between the two leaves. At some frequencies, the sound insulation may become inferior to that of a single wall due to resonance. Therefore, to approach an ideal condition for optimum sound insulation it is necessary to find out how to reduce the coupling through the structure and air.

In order to isolate the structural coupling, the leaves should be mounted on staggered studding with a resilient connection between stud and leaf. Also, in order to reduce the acoustic coupling through the air, the air space should be increased as much as possible and absorptive treatment introduced into the cavity between the leaves. Two typical examples are shown in Figure 6.16.

In the case of a fixed double- or triple-glazed window, since absorptive treatment in the air space is limited to the reveals as shown in Figure 6.17, it is recommended that the air space should be enlarged to increase the

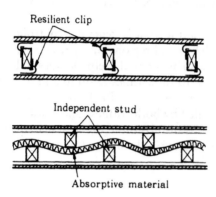

Resilient clip

Independent stud

Absorptive material

Figure 6.16 Typical double-leaf walls.

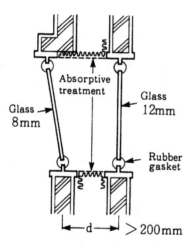

Absorptive treatment

Glass
8mm

Glass
12mm

Rubber gasket

d

> 200mm

Figure 6.17 Double-glazed window.

absorptive area and its absorption characteristic tuned to the resonance frequency of the air space. There are also techniques for reducing the effects of coincidence by using different thicknesses of glass for the two panes and attenuating well-defined resonances by slanting one pane with respect to the other. Ideally the panes should be mounted in resilient gaskets, completely isolated from the base structure, which consists of a double-leaf wall.

B. Theory of double-leaf walls

Let us consider the sound insulation when a plane wave is incident normally on two parallel, thin infinite walls separated by a distance d. As shown in Figure 6.18(a), the sound wave in the reverse direction in the air space

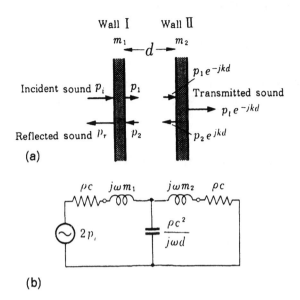

Figure 6.18 Sound insulation and analogous circuit for a double-leaf wall.

must be taken into account. For wall I, the equation of motion can be written

$$(p_i + p_r) - (p_1 + p_2) = Z_{w1} \left(\frac{(p_i - p_r)}{\rho c} \right) \tag{6.31a}$$

where Z_{w1} is the impedance per unit area of wall I.

Since there is continuity of particle velocity, we can also write

$$\frac{(p_i - p_r)}{\rho c} - \frac{(p_1 - p_2)}{\rho c} = 0 \therefore (p_i - p_r) - (p_1 - p_2) = 0 \tag{6.31b}$$

At wall II, there is a phase change due to the distance d, which leads to the following equation

$$\left(p_1 e^{-jkd} + p_2 e^{jkd} \right) - p_t e^{-jkd} = Z_{w2} \frac{p_t e^{-jkd}}{\rho c} \tag{6.32a}$$

and

$$\frac{\left(p_1 e^{-jkd} - p_2 e^{jkd} \right)}{\rho c} - \frac{p_t e^{-jkd}}{\rho c} = 0 \therefore \left(p_1 e^{-jkd} - p_2 e^{jkd} \right) - p_t e^{-jkd} = 0 \tag{6.32b}$$

where $k = \omega/c = 2\pi/\lambda$ and Z_{w2} is the impedance per unit area of wall II.

From the above four equations, the following can be derived

$$\frac{p_i}{p_t} = 1 + \frac{Z_{w1}+Z_{w2}}{2\rho c} + \frac{Z_{w1}\cdot Z_{w2}}{(2\rho c)^2}\left(1-e^{-jkd}\right) \tag{6.33}$$

Then, assuming that both walls have the same mass m per unit area for simplification, the impedance can be expressed from Equation (6.18) as

$$Z_{w1} = Z_{w2} = j\omega m$$

Substituting this into Equation (6.33), the transmission loss of the double-leaf wall is

$$R_{02} = 10\log_{10}\left|\frac{p_i}{p_t}\right|^2 = 10\log_{10}\left[1+4\left(\frac{\omega m}{2\rho c}\right)^2\left\{\cos kd - \left(\frac{\omega m}{2\rho c}\right)\sin kd\right\}^2\right] \tag{6.34}$$

When $d=0$,

$$R_{02} = 10\log_{10}\left[1+\frac{(2\omega m)^2}{(2\rho c)^2}\right] \tag{6.35}$$

which is the transmission loss of a wall of twice the mass of the single wall referred to in Equation (6.21). Also, when the expression

$$\left\{\cos kd - \left(\frac{\omega m}{2\rho c}\right)\sin kd\right\}^2 \tag{6.36}$$

becomes zero, then $R_{02}=0\,\mathrm{dB}$, i.e. there is no insulation. At low frequencies, if the wavelength λ is sufficiently large compared with the air space d, kd becomes small, hence

$$\frac{2\rho c}{\omega m} = \tan kd \approx kd = \frac{\omega}{c}d \therefore f_{rm} = \frac{1}{2\pi}\sqrt{\frac{2\rho c^2}{md}} \tag{6.37}$$

which means that at this frequency f_{rm} the sound insulation is zero.

f_{rm} can be thought of as the resonant frequency of a mechanical vibration system consisting of an air spring between two masses m. This can be represented by an analogous electrical circuit as shown in Figure 6.18(b), consisting of inductance, resistance and capacitance.

At high frequencies the expression (6.36) also becomes zero, i.e.

$$\frac{2\rho c}{\omega m} = \tan\frac{\omega}{c}d \tag{6.38}$$

for certain values of ω, and under these conditions $R_{02} = 0\,\text{dB}$. Although it is difficult to obtain this frequency f_{rd} analytically, it can be obtained graphically or numerically.

On the other hand, when

$$k_n d = (2n-1)\frac{\pi}{2} \quad (n=1,2,3,\ldots)$$ (6.39)

then expression (6.36) becomes

$$\left\{\cos kd - \left(\frac{\omega m}{2\rho c}\right)\sin kd\right\}^2 = \left(\frac{\omega m}{2\rho c}\right)^2$$

and R_{02} is a maximum value at frequencies given by

$$f'_{rd} = \frac{2n-1}{4}\cdot\frac{c}{d} \quad (n=1,2,3,\ldots)$$ (6.40)

Hence, Equation (6.34) can be written,

$$R_{02} \approx 40\log_{10}\left(\frac{\omega m}{2\rho c}\right) + 6$$ (6.41)

Thus, the R_{02} curve has many peaks and troughs, as shown in Figure 6.19.

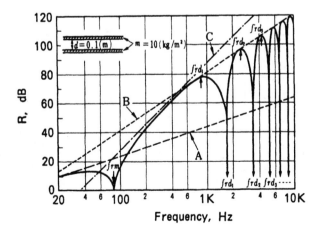

Figure 6.19 Theoretical values of double-leaf wall transmission loss. A: Equation (6.35), B: Equation (6.41), C: Equation (6.42) (Maekawa 1978).

The C line shows the following approximation derived from the analogous circuit in the frequency range between f_{rm} and f_{rd}

$$R_{02} = 10 \log_{10} \left(\frac{\omega^3 m^2 d}{2\rho^2 c^3} \right)^2 = 2R_{01} + 20 \log_{10} 2kd \qquad (6.42)$$

where R_{01} is the transmission loss of a single wall given by Equation (6.22).

C. *Practical sound insulation of double-leaf walls*

Some examples of the measured transmission loss of lightweight partitions, whose outer skins consist of thin panels, are shown in Figure 6.20. A dip may occur in the R curve around the resonance frequency f_{rm} (as given by Equation 6.37), beyond which the curve slopes up by about 10 dB oct^{-1} to higher frequencies. The peaks and dips due to f_{rd} in Equation (6.38) and due to f'_{rd} in Equation (6.40), may not appear distinctly, but the fall at 4 kHz coincides with the critical frequency f_c of a single plywood wall. Therefore, sometimes different panel thicknesses are employed so that the values of f_c for each panel are different.

There are two ways of increasing the R of this type of wall, as shown in the figure. One is by inserting absorbing material into the air space and the other is by underlining the outer skin with plasterboard. It can be seen that in (a) the effect of glass fibre is larger in the case of a lightweight partition, while in (b) inadequate application of plasterboard does not work so well.

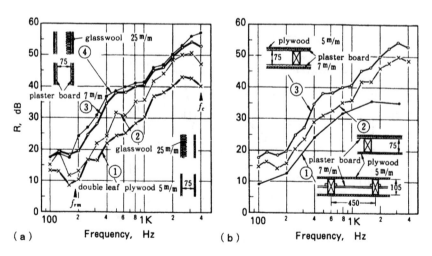

Figure 6.20 Improvement of transmission losses measured with lightweight partition walls (Maekawa 1968).

D. Sandwich panels

As an application of the above mentioned double-leaf wall construction, filling the air space with other materials, thus creating a sort of triple-layer wall, can be considered in order to increase thermal insulation or for other purposes. Theoretically, in place of the air spring previously discussed, the core material is thought of as acting as an elastic spring or resistance in the analogous electrical circuit. If the core material is glass fibre, for example, it acts as a resistance that improves the sound insulation, while an elastic material such as sponge or foamed plastic may produce more peaks and troughs in the frequency characteristics, thus decreasing sound insulation. Therefore, careful selection of material with reference to measured data is essential.

6.7 Effect of openings and cracks

A. Sound transmission through openings and cracks

(a) So far, the wall has been assumed to be airtight, but when a porous material or one with cracks is used, the transmission loss may be greatly reduced from the value estimated on the basis of the mass law, because sounds leak through the pores or cracks. For instance, a 10 cm thick bare concrete block, whose surface density is $160 \, \text{kg m}^{-2}$, has an average $R = 28$ dB, but when both faces are coated with oil paint, the R increases by as much as 13 dB up to 41 dB due to airtightness.

(b) If a fairly large opening is provided, the composite transmission loss (see Section 6.1A.d) should be calculated using a transmission coefficient $\tau = 1$ ($R = 0$ dB) for the opening area.

(c) Where a small opening exists, diffraction may occur. There is a complicated relationship between the wall thickness and the sound wavelength and in some cases $\tau > 1$ at the resonance frequency. Therefore, even with a tiny opening there might be an unexpectedly large sound transmission so that insulation may be seriously decreased. Thus, there must be careful detailing and construction in order to achieve sufficient airtightness at the partition panel joints and the perimeter around doors and windows.

Wilson & Soroka (1965) derived an approximate expression for the sound radiating from a circular hole in a wall. They assumed that sound incident on a cylindrical pipe of radius a and length l propagates as a plane wave inside the pipe and radiates sound from both ends, which vibrate like massless pistons. Figure 6.21 shows the result for $l = 30$ cm and $a = 5$ cm, compared with experimental values. At many resonance frequencies determined by the tube length (wall thickness), R becomes negative, which means

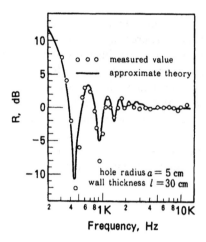

Figure 6.21 Example of transmission loss for circular hole (Wilson & Soroka 1965).

$\tau > 1$. Gomperts (1964) has derived another approximate formula by a different method. The resonance frequency is obtained when the condition $l + 2\delta = n\lambda/2$ $(n = 1, 2, 3, \ldots)$ is satisfied, where δ is the end correction and λ the wavelength. The maximum transmission coefficient $\tau_{max} = 2/(ka)^2$ where $k = 2\pi/\lambda$ at the resonance frequency. When $ka > \sqrt{2}$, $R = 0\,\mathrm{dB}$ is valid within $\pm 1\,\mathrm{dB}$.

Gomperts has also derived an approximate solution for a wall with slits, and gives the results compared with measured values (Figure 6.22). Theoretical values for an 11 mm wide slit in a wall are plotted between the measured values for 8 and 16 mm wide slits. Also, at the resonance frequency, it is found that $\tau > 1$ for the slit. The reason why peaks and troughs do not appear in the measured values is because measurements were made over wide frequency bands.

B. Sound insulation of windows and doors

The sound insulation of windows and doors depends not only on the insulation of the fittings but also upon the sealing around their perimeters.

The sound insulation of fittings is evaluated in the same way as for a single- or double-leaf wall, also taking into effect small cracks as discussed above. Generally, the insulation is governed by the latter so that detailing and accuracy of construction may produce very variable results. Thus, even though the R value of the door panel is increased, the insulation may not improve unless the gaps at the perimeter are minimised. Figure 6.23 shows two examples of details for the gap between door and floor, which is the weakest point. The first door has a gasket for sealing the edge of the door, and also absorptive treatment at the perimeter. It is essential to minimise

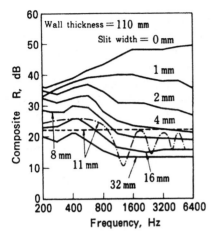

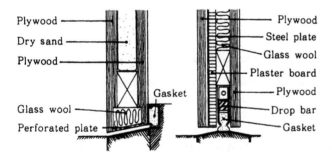

Figure 6.22 Example of measured composite transmission losses of walls with slits; wall size: 1.9 m × 1.9 m, slit located at centre; slit length: 1.9 m; solid line: measured value by 1/2 octave band; broken line: calculated value assuming slit R = 0; dash-dotted line: theoretical value of pure tone (Gomperts 1964).

Figure 6.23 Examples of sound-insulating doors.

the gap. The second door has a drop-bar, which automatically seals the gap when the door is in a closed position. This is very useful where a flat floor is required even under the door. In some cases, doors are weighted with dry sand or consist of dense boards and porous materials.

When an *R* value of more than 30 dB is required, it may be advantageous to employ a double-door arrangement with absorbent lining to the walls and ceiling between the doors as a sound-lock (see Section 6.6.A).

6.8   Flanking transmission

The sound transmission between two rooms in a building depends not only on the dividing construction, wall or floor, but the flanking building constructions can also give very important contributions to the transmission. Thus, between neighbouring rooms the apparent sound reduction index R', measured as described in Section 6.2, is typically some 2–5 dB less than the sound reduction index R of the partition wall itself. If the rooms are further apart and have no common wall, it is obvious that the sound transmission is entirely flanking transmission.

In some cases, flanking transmission can reduce the sound insulation seriously. Particularly dangerous constructions are lightweight facades or floors that continue from one room to the next; the sound generated in the source room creates vibrations in the flanking surface, and when these vibrations propagate to the next room they will radiate sound. There are three basic principles to avoid flanking transmission that will be discussed below.

A. Heavy flanking construction

In buildings with heavy constructions, like concrete or masonry, the flanking transmission is a minor problem if the flanking constructions are sufficiently heavy, i.e. having a surface mass at least 70–80% of that of the dividing construction. The heavy flanking construction means that the vibrations created by the incident sound have a very small velocity and, thus, the sound energy that can be transmitted to the next room is very limited.

B. Separation of flanking constructions

The best way to avoid flanking transmission is often to disconnect the flanking structure at the point where it passes the dividing construction. However, in practice it is not always so easy, and some kind of resilient layer can be the solution instead of a gap. Such solutions may put high demands on the materials and on the correct execution. Some examples are shown in Figure 6.24.

C. Treatment to reduce sound radiation

The vibrations generated in the flanking construction may not necessarily radiate much sound energy if the construction is a panel with a sufficiently high critical frequency, as will be discussed later in Section 7.1. A gypsum board with thickness 13 mm or less is a typical example of such a panel. At frequencies below the critical frequency the vibrations in the panel are travelling at a lower speed than the sound in the air, and thus the sound radiation is not efficient. Some examples are shown in Figure 6.25.

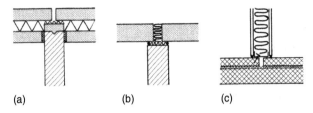

Figure 6.24 Examples of separation to avoid flanking transmission. (a) Concrete wall connected to a facade of sandwich-elements. (b) Concrete wall connected to ceiling elements of lightweight concrete; elastic sealing and mineral wool in the gaps. (c) Floating floor with a gap under a light partition wall.

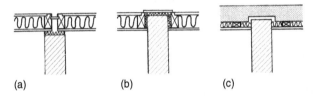

Figure 6.25 Examples of flanking constructions with reduced sound radiation. (a) Light wooden facade elements connected to a massive partition wall. (b) Same as (a) but improved to avoid leakage problems. (c) Lightweight concrete facade with internal treatment of 9 mm gypsum board on laths.

6.9 Noise reduction in air ducts

Ducts convey not only air but also noise efficiently. Noise reduction means the prevention of sound transmission whilst conveying air. We must start by analysing sound transmission through ducts.

A. Natural attenuation of sound waves in tubes

When a plane wave propagates along a tube whose cross-sectional area is S, from Equations (1.10) and (1.12) the acoustic impedance Z_A is

$$Z_A = \frac{p}{Sv} = \frac{\rho c}{S} \tag{6.43}$$

Since Sv is the volume velocity, Z_A in this case is called the volume impedance, and is inversely proportional to the cross-sectional area.

a. Attenuation due to changes in cross-section

When an acoustic tube section abruptly changes, as shown in Figure 6.26, since the impedance also changes the plane wave propagating in the tube

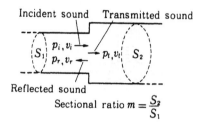

Sectional ratio $m = \dfrac{S_2}{S_1}$

Figure 6.26 Cross-sectional area change in acoustic tube.

loses a portion of its energy by reflection, in the same way as sound loses its energy at the boundary plane between two media in Figure 1.8.

Both result in sound attenuation. Equations (1.28)–(1.32) are used to calculate the attenuation but now with volume impedances in place of the characteristic impedance.

i.e. $Z_1 = \dfrac{\rho c}{S_1}, \quad Z_2 = \dfrac{\rho c}{S_2}$

Hence, introducing the cross-sectional ratio $m = S_1/S_2$,

$$
\begin{aligned}
R &= 10\log_{10}\frac{|p_i|^2\,S_1}{|p_t|^2\,S_2} = 10\log_{10}\left(\frac{S_1+S_2}{2S_1}\right)\frac{S_1}{S_2}\\
&= 10\log_{10}\frac{1}{4}\left(m+\frac{1}{m}+2\right)(\text{dB})
\end{aligned}
\tag{6.44}
$$

This value is valid when both tubes are infinitely long, i.e. they have non-reflective ends. Actually, at the junction, the attenuation may be more complicated depending on the connecting impedances, but it is generally considered that attenuation occurs at any discontinuous sections.

b. Attenuation due to reflection at duct open end

An open end of a duct, such as an air outlet or inlet, is such an abrupt change of cross-section that the attenuation is large. The sound waves outside the open end may be thought of as radiated by an imaginary piston vibrating at the open end of the duct instead of as plane waves. The results from the theoretical derivation of the **radiation impedance*** of the piston are shown in

---

*Radiation impedance: when a vibrating plate radiates a sound wave with vibrating velocity v, the ratio of it to the reaction F_r of the vibrating plate is given by $Z_r = v/F_r$. Z_r is called the radiation impedance. F_r is obtained by integration of the reactions of all portions of the vibrating plate (see Lit. B7).

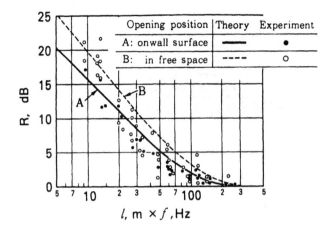

Figure 6.27 Attenuation due to reflection at duct open ends (open-end area l^2).

Figure 6.27, while the measured values were obtained by experiments that arranged for the duct to have an open end in a reverberant room (Maekawa *et al.* 1971).

c. Attenuation due to duct bends

If the duct has a radius of curvature at a bend, attenuation may be small, but at a right-angled bend the attenuation is obtained as shown in Figure 6.28, because of phase differences between inner and outer paths. When it is necessary to avoid increasing the air-flow resistance, the internal side should be curved while the external side should be right-angled.

The application of an absorptive lining to the inside of a bend is very effective for noise reduction, where additional attenuation can be expected by an amount shown by the dotted line in Figure 6.28, in addition to the attenuation caused by the lining calculated by Equation (6.45) in the next section.

B. Attenuation devices in ducts

There are two types of device used to increase noise reduction in ducts; one device is dissipative and dissipates sound energy by absorption, while the other is reactive and reflects sound back to the source due to an abrupt change in impedance. Figure 6.29 shows their conceptual designs.

C. Lined ducts

The energy flowing in a duct is proportional to the cross-sectional area S (m²) and the amount absorbed is proportional to the perimeter P (m).

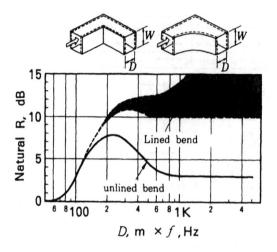

Figure 6.28 Natural attenuation in right-angled duct bend, lining length should be at least 2D to 4D beyond the bend. W is arbitrary.

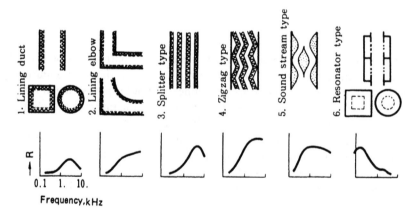

Figure 6.29 Typical attenuation devices and their schematic characteristics.

Therefore, the attenuation R is approximately given by

$$R = K \cdot \frac{P}{S} (\text{dB m}^{-1}) \tag{6.45}$$

where K is a constant that is determined by the absorption coefficient of the lining material, as shown in Figure 6.30.

Equation (6.45) is valid up to the frequency at which the wavelength is equal to the diameter or the short-side width of the duct, beyond which the

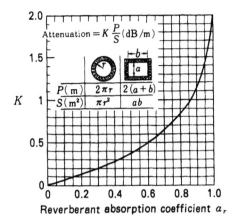

Figure 6.30 Values of K in Equation (6.45) for absorption coefficient of lining material.

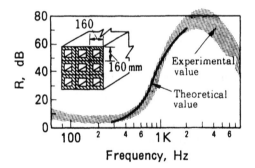

Figure 6.31 Example of a cell-type lined duct; rock wool lining (180–200 kg m^{-3}), 25 mm thick; solid line: calculated value by Equation (6.45) (Maekawa 1957).

sound wave becomes a beam flowing along the centre of the duct, in which case the attenuation decreases.

Figure 6.31 shows an example of a multi-cell duct that has in it many small sections, so as to increase P/S without reducing the volume of air flow. The figure shows that measured values decrease at frequencies above 3000 Hz, at which point the wavelength is equal to the internal size of the cell.

These characteristics of lined ducts can be fully analysed with the aid of wave theory in closed spaces (see Lit. B6, B7). However, these discussions seem to be too complicated for practical application.

The parallel baffle, type 3 of Figure 6.29, which is a variation of the lined duct, has also the disadvantage of decreased attenuation at the high

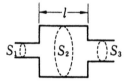

Figure 6.32 Expansion-type muffler.

frequencies. The zigzag, type 4, is an improvement but its resistance to air flow is larger. In order to reduce such resistance, the sound stream, type 5, has been devised. They are available on the market, ready fabricated in various sizes.

In the high-frequency range, the simplest method of increasing effective attenuation is to line the right-angled bends, in such a way as to increase the number of lined elbows until sufficient attenuation is obtained using Figure 6.28 and Equation (6.45).

D. Expansion-type mufflers

When the cross-section of a portion of a duct is expanded to form a cavity, as shown in Figure 6.32, the latter becomes an acoustic filter due to the cross-sectional changes at two places. The attenuation for plane wave propagation may be derived by the same principle used in Figures 6.26 and 6.18. In the simplest case, when $S_1 = S_3$, with $m = S_2/S_1$ and $k = 2\pi/\lambda$

$$R = 10\log_{10}\left\{1 + \frac{1}{4}\left(m + \frac{1}{m}\right)^2 \sin^2 kl\right\} \text{ (dB)} \qquad (6.46)$$

is obtained (see Lit. B14a).

Examples are shown in Figure 6.33. As m is increased, the peak of the attenuation becomes higher, and when l is smaller, the number of troughs is reduced, thus, improving the performance. Although Equation (6.46) shows the attenuation characteristics, it should be noted that every dimension of the cavity should be smaller than a wavelength in order to satisfy the condition of plane wave propagation in the duct. If the frequency becomes higher, this condition may not hold.

E. Sound-absorbing chambers

When an air chamber is provided at the junction of ducts or at the base of outlets, it acts as an expansion muffler. However, at any frequency for which the wavelength is smaller than the dimension of the chamber, Equation (6.46) does not apply. In practice, absorptive treatment is required inside

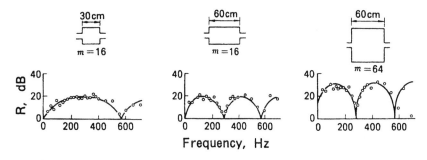

Figure 6.33 Examples of attenuation characteristics of expansion-type mufflers (D.D. Davis Lit. B14).

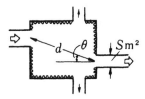

Figure 6.34 Sound-absorbing chamber.

the chamber so that geometrical acoustics may have to be used. Hence, referring to Figure 6.34, Equation (6.47) is obtained:

$$R = 10 \log_{10} \frac{1}{S \left[\dfrac{\cos \theta}{2 \pi d^2} + \dfrac{1 - \alpha}{A} \right]} \text{(dB)} \tag{6.47}$$

where S is the outlet area from the chamber (m^2), d is the distance from inlet to outlet (m), θ is the angle between d and the normal to the outlet surface, A is the total absorption in the chamber (m^2) and α is the absorption coefficient of the lining material.

The first term of the denominator of the equation gives the direct sound energy from inlet to outlet, while the second term gives the diffused sound to the outlet. If some baffles are provided inside, so that the direct sound cannot reach the outlet, the following approximation may be used

$$R = 10 \log_{10} \frac{A}{S (1 - \alpha)} \text{(dB)} \tag{6.48}$$

Since in the derivation of these equations, a diffuse sound field is assumed, the approximation can hold well into the higher frequency ranges where the wavelength is smaller than the chamber dimension.

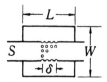

Figure 6.35 Resonator-type muffler.

F. Resonator-type mufflers

The attenuators so far discussed, particularly the dissipative type, are generally effective at high frequencies, but poor at low frequencies. In order to improve the attenuation at low frequencies, resonator attenuators are often used. These consist of N perforations and an exterior airtight cavity, as shown in Figure 6.35. Hence, the resonance frequency from Equation (4.21) is

$$f_r = \frac{c}{2\pi} \sqrt{\frac{NG}{V}} \, \text{(Hz)} \tag{6.49}$$

where G is the conductivity for each perforation, i.e. $G = s/(l + 0.8\,d)$; l is the wall thickness, d is the hole diameter and s is the area of a hole. Then, the attenuation is given by

$$R = 10 \log_{10} \left[1 + \left\{ \frac{\sqrt{NGV}}{(f/f_r) - (f_r/f)} \right\}^2 \right] \text{(dB)} \tag{6.50}$$

which shows very large attenuation around the frequency f_r. However, since the octave-band attenuation should be taken into account for rating with noise criteria, f_r should be taken as the centre frequency, then

$$\text{octave band attenuation} = 10 \log_{10} \left[1 + \frac{NGV}{2S^2} \right] = 10 \log_{10} \left[1 + 2K^2 \right] \text{(dB)} \tag{6.51}$$

where

$$K = \frac{\sqrt{NGV}}{2S} \tag{6.52}$$

For this calculation, Figure 6.36 is very useful (Maekawa 1959).

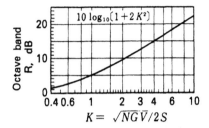

Figure 6.36 Octave-band attenuation of resonator-type muffler.

[Ex. 6.3] Let us design a resonator-type attenuator for which the octave-band attenuation at the mid-frequency 70 Hz, must be 10 dB in a duct of diameter 180 mm.

From Figure 6.36, $K = 2$ and the duct cross-sectional area

$$S = 90^2 \ \pi = 20{\cdot}4 \times 10^3 \text{ mm}^2$$

From Equation (6.52)

$$\sqrt{NGV} - 2 \times 2S = 102 \times 10^3 \text{ mm}^2$$

From Equation (6.49)

$$\sqrt{\frac{NG}{V}} = \frac{70 \times 2\pi}{340 \times 10^3} = 1{\cdot}31 \times 10^{-3} \text{ mm}^{-1}$$

Hence,

$$V = \frac{\sqrt{NGV}}{\sqrt{\dfrac{NG}{V}}} = 78 \times 10^6 \text{ mm}^3$$

$$NG = 1{\cdot}31^2 \times 78 = 132 \text{ mm}$$

For a hole with $d = 12$ mm, $l = 0.7$ mm, then $G = 11.0$ mm.

$$N = \frac{132}{11} = 12 \text{ pcs}$$

For the determination of the dimensions shown in Figure 6.35, the following conditions must be satisfied.

Condition 1: W or $L < \lambda_r/3$ where λ_r is the wavelength at the resonance frequency. Therefore, the design cannot be used for high frequencies.

Condition 2: The range of perforation $\delta < \lambda_r/12$ must be concentrated at the centre of the cavity, while the pitch of holes should be about $2d$ otherwise G will be changed.

PROBLEMS 6

1　When a plane sound wave is incident at angle θ on a single wall, as shown in Figure 6.7, derive Equation (6.23) for the sound transmission loss. Verify Equation (6.24) for the random incidence mass law of sound transmission loss of a single wall.

2　When a plane sound wave is incident at an angle θ on a double-leaf wall, as shown in Figure 6.18, derive an equation expressing the sound transmission loss of the wall.

3　Explain the basic principles necessary to improve the sound insulation of lightweight walls.

4　Outline the thoughts required on architectural planning and design in order to achieve good sound insulation in a window that is required for good natural lighting.

5　Enumerate the principal devices for noise attenuation in the noise control design of an air-conditioning system.

6　Design a muffler to provide 15 dB attenuation in the 1/3 octave band at 63 Hz in an air duct (cross-sectional area 400 m × 400 m). Then, sketch the frequency characteristics of the muffler's sound attenuation by theoretical calculations.

7 Isolation of structure-borne noise and vibration

Structure-borne sound that is transmitted through solid material is essentially a vibration of the material. In this chapter the fundamentals of the isolation of structure-borne sound and vibration are discussed.

7.1 Propagation and radiation of structure-borne sound

A. *Generation of structure-borne sound*

When impacts resulting from footsteps, the slamming of doors, furniture movement, etc. are transferred to the building structure, there is a resulting vibration that is propagated through the structure, known as structure-borne sound.

Vibration of mechanical equipment in the building such as a pump, blower, elevator, refrigerator, etc. generates structure-borne sound as steady-state sources.

Plumbing and piping for steam or water, for example, also generate structure-borne sounds at taps and valves and convey those intermittent structure-borne sounds through themselves and the building structure.

Sound generated in a room excites the walls, ceiling and floor into vibration, which is also propagated as sound-induced structure-borne sound.

In addition, noise and vibration generated by traffic, construction work or industry outside the building is transmitted into the building through the ground and the building foundation as structure-borne sound as shown in Figure 7.1.

Structure-borne sound is perceived by a listener as a consequence of the airborne sound radiated from the vibrating surfaces.

In order to quantitatively predict the generation of structure-borne sound, we have to calculate the vibration velocity or acceleration that occurs in the structure from a knowledge of the exciting force of impact or driving force of vibration that causes the structure-borne sound, and, hence, obtain the mechanical impedance at the receiving point. A design method of vibration

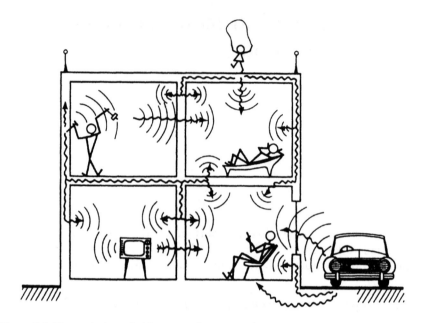

Figure 7.1 Transmission of airborne and structure-borne sound.

isolation of a machinery base using an analogous electrical circuit is given in the literature (Breeuwer & Tukker 1976).

Applying this method, the structure-borne sound can be calculated. Unfortunately there are few quantitative data on the exciting force and other parameters that are used for the calculation and, moreover, it is not easy to measure them. Therefore, more effort is needed to obtain the necessary data so that the process of predicting structure-borne sound is the same as for airborne sound.

B. *Propagation and measurement of structure-borne sound*

a. *Wave types in solids*

Although sound in air is propagated only as longitudinal waves, in solids it may be transmitted by several types of wave, i.e. a longitudinal (compressional) wave, a transverse (shear) wave, etc., since solids have not only compressional stiffness but shear stiffness as well. They often combine with each other to make a more complicated wave field, such as a bending wave or surface (Rayleigh) wave. In some cases the wave speed depends on frequency, thus giving rise to dispersion. The motion of elements of the solid material and formulas for the wave speed for the various wave types are shown in Figure 7.2.

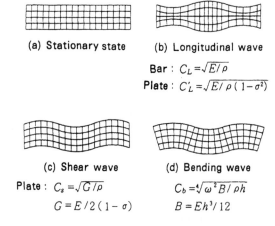

(a) Stationary state (b) Longitudinal wave

Bar : $C_L = \sqrt{E/\rho}$

Plate : $C_L' = \sqrt{E/\rho(1-\sigma^2)}$

(c) Shear wave (d) Bending wave

Plate : $C_s = \sqrt{G/\rho}$ $\qquad C_b = \sqrt[4]{\omega^2 B/\rho h}$

$\qquad G = E/2(1-\sigma)$ $\qquad\quad B = Eh^3/12$

Figure 7.2 Wave types and speeds of sound in solids (Beranek 1954, Lit. B13).
E: Young's modulus, ρ: density of the material, kg m^{-3}; σ: Poisson's ratio;
G: shear modulus; B: bending stiffness per unit width; and h: thickness.

Generally, the speed of longitudinal waves in a solid is very high, as shown in Table 1.1 in Chapter 1, and attenuation very small. The bending wave speed varies with frequency, as shown in Figure 6.11, resulting in the coincidence effect.

b. Measurement of structure-borne sound

A vibration pickup, as in Table 2.8, is mounted rigidly on a vibrating solid surface and detects the vibration perpendicular to the surface.

In order to eliminate the effect of the pickup itself, a very small and light piezoelectric sensor is often used. The data are given in terms of the velocity level L_v, which is suitable for comparing with airborne sound data.

$$L_v = 10 \log_{10} \frac{v^2}{v_0^2} \qquad\qquad (7.1a)$$

where v^2 is the mean square velocity, and v_0 is the reference value.

Although the ISO 1683 Standard recommends $v_0 = 10^{-9}$ m s^{-1}, in this text we use

$$v_0 = 5 \times 10^{-8} \ (\text{m s}^{-1}) \qquad\qquad (7.1b)$$

because it is more convenient for calculating the sound radiation as described in the next section.

Since v is related to the acceleration a (m s^{-2}) where $a = 2\pi f v$, L_v can also be related to the acceleration level L_a as follows

$$\frac{v}{v_0} = \frac{a}{a_0} \cdot \frac{a_0}{2\pi f v_0}$$

$$\therefore \quad L_v = L_a + 20\log_{10}\frac{a_0}{(2\pi f v_0)} \qquad (7.2)$$

where a_0 is the arbitrary reference value for L_a. (The reference $a_0 = 10^{-6}$ m s^{-2} is recommended by ISO 1683.)

Figure 7.3 shows some examples of measured velocity levels (Lit. B34).

C. Sound radiation from a vibrating solid body

a. Sound radiation ratio of vibrating piston

When an infinite plane rigid wall vibrates as a piston with velocity v, the air particles in contact with the wall surface also vibrate with velocity v, so the sound pressure is $\rho c v$ since the impedance of the air is ρc. Consequently, the radiated sound power is $\rho c v^2$ per unit area. Then, the sound power level is

$$L_W = 10\log_{10}\frac{\rho c v^2}{10^{-12}} = 10\log_{10}\frac{v^2}{v_0^2} + 10\log_{10}\frac{\rho c v_0^2}{10^{-12}} \qquad (7.3)$$

substituting $v_0 = 5 \times 10^{-8}$ m s^{-1}, the second term can be neglected, then $L_W = L_v$, and the velocity level is the sound radiation power level itself.

When the size of the plate is less than the wavelength of the sound in air, the air particles adjacent to the plate surface move into the surroundings

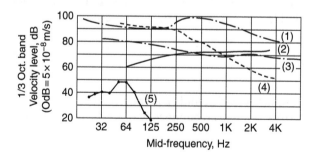

Figure 7.3 Examples of velocity levels per 1/3 octave (M. Heckl Lit. B19): (1) subway rail when train is passing at 60 km h^{-1}; (2) standard tapping machine acting on a 12-cm concrete floor; (3) electric motor on elastic mounts running at 1400 rpm; (4) elastically mounted elevator for six persons; and (5) house wall when a street-car is passing at 45 km h^{-1} and at 13 m distance.

or to the rear side, so that pressure changes do not occur. This fact means that the sound radiation efficiency varies with the relative sizes of the plate and wavelength. Then, the sound power W radiated from a plate of area S vibrating with velocity v, is expressed as

$$W = \sigma_{\text{rad}} \rho c v^2 S \qquad (7.4a)$$

where σ_{rad} is called the **radiation ratio.**
  Logarithmically, we can write

$$10 \log_{10} \sigma_{\text{rad}} = L_W - (L_v + 10 \log S)\,(\text{dB}) \qquad (7.4b)$$

This is the difference between the power level and the velocity level when $S = 1\,\text{m}^2$.
  As the most basic condition, the radiation ratio of a circular or rectangular piston, set in a sufficiently large rigid wall, corresponds to the real part of the radiation impedance (see Section 6.9A.b). Its theoretical values are shown in Figure 7.4. At higher frequencies, where the wavelength is smaller than the piston diameter, the value of σ_{rad} becomes unity, and the velocity level L_v becomes equal to the power level L_W per unit area, but at lower frequencies, at which the wavelength is larger than the diameter, the value decreases by 6 dB oct^{-1}.

b. Sound radiation ratio for bending vibration

If a wide wall is set into free bending vibration by any means, as shown in Figure 6.11, it radiates a plane wave in the direction θ which satisfies the relation

$$\sin \theta = c/c_B \qquad (7.5)$$

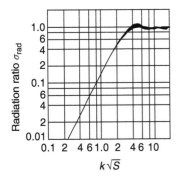

Figure 7.4 Radiation ratio of a rectangular piston, $k = 2\pi/\lambda$, S: area of a rectangular piston whose aspect ratio is less than 2, and also an approximation for a circular piston.

from Equation (6.28). Therefore, when $c < c_B$, i.e. $f > f_c$ for frequencies higher than the **critical frequency**, the radiation ratio $\sigma_{rad} = 1$.

When $f = f_c$, σ_{rad} is somewhat larger than one, and when $f > f_c$ the direction which satisfies Equation (7.5) does not exist. Then, even if the air particles in contact with the surface are moved by bending vibration of the wall, they do not create any pressure change. That is, the radiation ratio is very much decreased and its value depends on the value of f/f_c and of the internal energy losses.

In a finite wall, because it has resonant vibration modes as shown in Figure 4.18 and Equation (4.20a), the radiation behaviour is very complicated. But the approximate method of obtaining the average radiation ratio vs frequency for a rectangular panel is obtained by using the space-time average mean-square velocity of the panel, as quoted in the literature (see Lit. B22b, p. 294). Cremer and Heckl (1976) (Lit. B19a) presented theoretical results and compared them with experimental results as shown in Figure 7.5 (see Lit. B.19b, p. 496).

c. Room noise level resulting from the vibration of the enclosure

If a part of the enclosure S_i (m²) vibrating with velocity v_i (m s⁻¹) radiates sound into the room, then when every part of the enclosure vibrates the

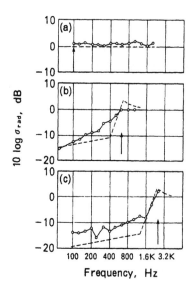

Figure 7.5 Radiation ratios of lightly damped point-excited plates (dashed curves are theoretical ones) (Cremer *et al.* Lit. B19): (a) 24-cm thick brick wall of 12 m² area; (b) 7-cm thick wall of light concrete, 4 m² area; and (c) 13-mm plasterboard wall, divided into 0.8 m² panels by lath grid work.

total sound power radiates $W = \Sigma\, I_i\, S_i$. The sound energy density is then, from Equations (3.23) and (7.4a)

$$E = \frac{4\rho c \sum \sigma_{\mathrm{rad}\,i} v_i^2 S_i}{cA} \tag{7.6}$$

The value of the radiation ratio in the frequency range higher than the critical frequency can be approximated by $\sigma_{\mathrm{rad}} = 1.0\text{–}1.1$, however, at lower frequencies the estimation from Figure 7.5 proves rather difficult.

7.2 Reduction of structure-borne noise

Structure-borne sound waves are reflected and attenuated during propagation more or less at every discontinuity, such as at a sudden change in cross-section or material, due to changes in direction at bends or branches, as well as due to added masses (see Lit. B.19b, Chap. V).

A. Reduction by change of cross-section

A longitudinal wave is attenuated at a sudden change of cross-section, as shown in Figure 6.26, Equation (6.44) is valid in this case.

For bending waves, four boundary conditions for continuity must be satisfied at $x = 0$. At the point of change, as shown in Figure 7.6(a)

1	velocities	$V_{y1} = V_{y2}$	(7.7a)
2	shear forces	$F_{y1} = F_{y2}$	(7.7b)
3	angular frequencies	$\omega_{z1} = \omega_{z2}$	(7.7c)
4	moments	$M_{z1} = M_{z2}$	(7.7d)

For the incident wave propagating from the left $(x < 0)$, the first boundary condition (7.7a) leads to the simple expression at $x = 0$.

$$1 + r + rj = t + tj \tag{7.8}$$

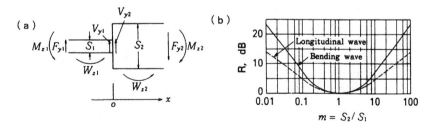

Figure 7.6 Transmission loss at a discontinuity in cross-section, as a function of the thickness ratio (Cremer *et al.* Lit. B19).

where r is the reflection coefficient, t is the transmission coefficient and rj, tj are coefficients that express the near-field vibration decaying far from $x = 0$. At $x = 0$, the four unknowns may be complex quantities to be determined from the four boundary conditions; t is of greatest interest here. Since the process is complicated, the results are only presented here as a transmission loss

$$R_B = 10 \log_{10} \left(\frac{X^{5/4} + X^{3/4}}{1 + X^2/2 + X^{1/2}} \right)^2 \text{ (dB)} \tag{7.9}$$

where $X = m + m^{-1}$ and $m = S_2/S_1$.

These values are shown in Figure 7.6(b), together with the values for the longitudinal wave. We find that a cross-sectional change does not give a large reduction of structure-borne noise in building practice.

B. Reduction by bends and branches at right angles

When a free bending wave reaches a bend or a branch at a right angle in a bar or plate, the behaviour of reflection and transmission at these junctions is complicated as some of the partial wave changes its wave type. Some examples of the results of theoretical analysis are shown in Figures 7.7 and 7.8.

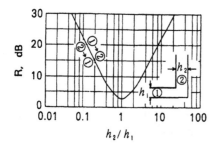

Figure 7.7 Transmission loss for bending waves at corners (Cremer *et al.* Lit. B19).

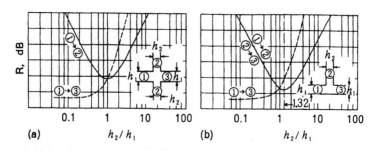

Figure 7.8 Transmission loss for bending waves (Cremer *et al.* Lit. B19). (a) At plate intersections; and (b) at plate branch.

Further reductions may be obtained by adding masses en route or at bends and branches, such as ribs, beams or columns, which are called **blocking masses**. Also, with random incidence at the junctions, a little more reduction is possible, because with oblique incidence the reduction increases with the angle of incidence.

C. Reduction by different materials

a. Reduction at the boundary between two materials

The behaviour of a longitudinal wave incident normally on a boundary between two different materials, has already been illustrated in Figure 1.8 (Chapter 1). By using the notation of Figure 1.8, the transmission loss is

$$R = 10 \log \frac{p_i v_i}{p_t v_t} = 10 \log \frac{1}{t_p \cdot t_v} = 10 \log \frac{(Z_1 + Z_2)^2}{4 Z_1 Z_2} \text{(dB)} \qquad (7.10)$$

b. Reduction by resilient layers

Resilient layers, such as rubber, springs, etc., are efficient means of reducing structure-borne sound. Figure 7.9 shows analytical results produced by Cremer *et al.* (Lit. B19). The softer the resilient materials the more effective they are. However, when the interlayer thickness becomes large, the phase shift in the material has to be taken into account, as the total transmission may occur in the important frequency region, a phenomenon similar to that illustrated by Equation (6.46) and Figure 6.33 (Chapter 6).

As for bending waves, the behaviour is complicated by deformation of the resilient material. The total transmission can be found at the particular frequency 170 Hz, in Figure 7.9.

D. Reduction of structure-borne noise in a real building

A real building is a complicated combination of plates and bars that form the construction of walls or floors and columns or beams, respectively. It

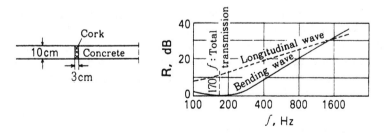

Figure 7.9 Transmission loss of an elastic interlayer (Cremer *et al.* Lit. B19).

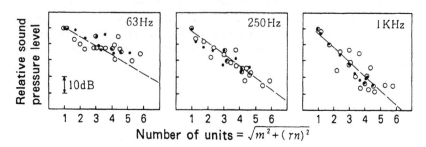

Figure 7.10 Attenuation of structure-borne sound measured in an existing 10 storey building. The unit is $(6.0 \times 5.9)\,m^2 \times 2.9\,m$ high, m and n are the number of rooms counted horizontally and vertically, respectively, γ is a proper weighting factor. The circles o and ● are measured on the floors above and under the excited wall, respectively. Dashed lines are calculated by the SEA method.

is so difficult to deal with structure-borne noise analytically that statistical energy analysis is often used (see Section 11.15). However, there are still many problems left in attempting to obtain a precise estimation of structure-borne noise.

Figure 7.10 shows an example of measured structure-borne noise levels in rooms caused by the vibration excited by a concrete chipping machine in an existing hospital, which has a rather simple construction of reinforced concrete, compared with values calculated by the method of SEA (Furukawa *et al.* 1990).

7.3 Measurement and rating of impact sound insulation

The amount of insulation provided by building elements against structure-borne noise is only measured and classified for floors and ceilings. A standard impact machine specified by ISO 140-6 is used for this purpose as follows.

A. Laboratory measurement of impact sound insulation of floors

The test specimen is installed in the test opening between two reverberant rooms, No. 2 and No. 3, shown in Figure 6.4. The space and time average sound pressure levels in the receiving room, when the test floor is excited by the standard tapping machine, are measured and denoted by L_i, called the impact sound pressure level. The **normalised impact sound level** is defined in the same way as in Equation (6.14).

$$L_n = L_i + 10 \log_{10} (A/10)\ (dB) \qquad (7.11)$$

where A is the sound absorption area obtained from the measured reverberation time and using Equation (3.26).

ISO 140-6 specifies the details as follows:

1 The tapping machine should have 5 hammers, each of weight 0.5 kg, placed in line, with 40 cm between both ends. The hammers are freely dropped from 4 cm high onto the specimen successively, at a rate of 10 times per second.
2 The size of the test specimen should be between 10 and 20 m² with the shorter edge length not less than 2.3 m.
3 The tapping machine should be placed in at least four positions. In the case of an anisotropic floor construction (ribs, beams, etc.) more positions are necessary. In addition, the hammer connection line should be orientated at 45° to the direction of the beams or ribs. The distance of the tapping machine from the edges of the floor should be at least 0.5 m.
4 The impact sound pressure level in the receiving room should be averaged using Equation (1.22), with measured values at a number of microphone positions, or alternatively by means of a continuously moving microphone. It is recommended that a Class 1 sound level meter is used with time weighting S, as shown in Table 2.2.
5 The sound pressure level should be measured using 1/3 octave or octave-band filters of which the frequency range should be at least from 100 to 3150 Hz (preferably 4000 Hz) for 1/3 octave bands, and from 125 to 2000 Hz for octave bands.

B. Field measurement of impact sound insulation of floors

When the purpose of the field measurements is to determine the impact sound insulation properties of a building element, the **normalised impact sound level** (Equation (7.11)) and the **standardised impact sound level**, normalised by a reference value of reverberation time 0.5 s, in the same manner as Equation (6.15), are used. However, in these cases they should be denoted by L'_n, and $L'_{n,0.5}$, respectively, because there is flanking transmission.

When evaluating the effect on the occupants of the buildings, it is considered that the impact sound pressure level, L'_i, itself without any normalisation, is appropriate, though this is in contradiction to ISO 140-7.

In Japan there is a standard, not only according to this concept, but using a heavy weight impact machine simulating children's jumping, which is used in addition to the standard tapping machine (see Section 11.12.D).

C. Single number rating of impact sound insulation of floors

As for airborne sound insulation, a method of single number rating for impact sound insulation of floors is formalised by ISO 717-2 as follows.

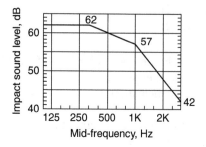

Figure 7.11 Curve of reference value for floor impact noise (ISO 717-2).

1 The reference curve, shown in Figure 7.11, is shifted towards the measured curve of L_n (or L'_n, and/or $L'_{n,0.5}$) values until the mean unfavourable deviation is as large as possible, but not more than 2.0 dB.
2 The value of the reference curve at 500 Hz, L_{nw} (or L'_{nw}), is called the weighted normalised impact sound pressure level, and $L'_{n,0.5w}$ is called the weighted standardised impact sound pressure level.

There are two problems, one is the inadequacy of the standard tapping machine for simulating the impacts occurring in a dwelling, and the other is the long time it takes to make the field measurements specified by ISO. In Section 11.12, other methods are given, which are used in many countries with their own standards or codes and criteria.

D. Reduction of the impact sound transmission through floors

Generally, it is so difficult to block the transmission of structure-borne sound, that we must do our best to suppress the generation of the shock or vibration. For this purpose, soft or resilient material should be used on the floor finish or on any other areas where impact sound may occur.

The reduction of the impact sound transmission through a floor is said to be 5–10 dB with a cork finish, and 4–20 dB with a carpet. Figure 7.12 shows examples of the effect of various finishes measured with a standard tapping machine, compared with a bare concrete floor, using the method specified in ISO 140-8. It can be seen that softer materials, even if thin, give a greater reduction at higher frequencies. Furthermore, thicker materials, such as the Japanese *tatami* (a thick straw mat), give a greater reduction of the impact sound, even at low frequencies.

The effect of finishes on the impact of a heavy load is so small that the mass and stiffness of the floor slab need to be increased to deal with it. So,

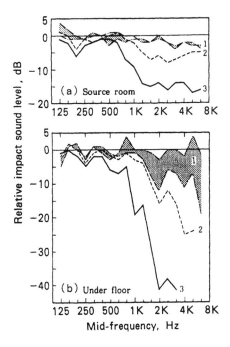

Figure 7.12 Reduction of impact sound from floor by means of various finishes: 1: vinyl tile; 2: linoleum; and 3: vinyl sheet on foam underlay 3 mm thick; (a) reduction of impact sound pressure level in the source room; and (b) reduction of impact sound in the receiving room under the excited floor (Maekawa 1965).

a concrete floor should have a thickness of more than 20 cm depending on floor area.

It is essential that any impact should not be transmitted directly to the building structure, and ideally, not only the floor but also the ceiling and walls should be floated from the main structure using the principle of vibration damping discussed in the next section.

7.4   Principle of vibration isolation

In order to deal precisely with the vibration of a solid body, vibrations in each direction of a three-dimensional coordinate axis, around which rotational vibrations can also take place, need to be considered. This results in a requirement for analysis of 6 d.f. (degree of freedom). However, the principle of vibration damping is presented here on the basis of a simple vibration system based on 1 d.f., as shown in Figure 7.13, which still has application in building practice.

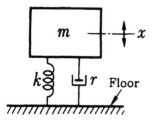

Figure 7.13 Principles of vibration isolation.

A. Vibration of a 1-degree-of-freedom system

A body whose mass is m rests on an elastic spring, which has elastic modulus k. When it is forced to vibrate in the vertical direction by an external force $P\cos\omega t$, its motion can be represented by

$$m\frac{d^2x}{dt^2} + r\frac{dx}{dt} + kx = P\cos\omega t \qquad (7.12)$$

where r is the frictional resistance factor.

If the vibrational driving force to the floor is P_t

$$P_t = kx + r\frac{dx}{dt} \qquad (7.13)$$

From Equations (7.12) and (7.13), the vibration transmissibility T is derived as follows,

$$T = \frac{|P_t|}{|P|} = \left\{ \frac{1 + \left(2\cdot\dfrac{\omega}{\omega_n}\cdot\dfrac{r}{r_c}\right)^2}{\left(1 - \dfrac{\omega^2}{\omega_n^2}\right)^2 + \left(2\cdot\dfrac{\omega}{\omega_n}\cdot\dfrac{r}{r_c}\right)^2} \right\}^{1/2} \qquad (7.14)$$

where $\omega_n = 2\pi f_n = \sqrt{k/m}$, f_n is the natural frequency, and the critical damping resistance, $r_c = 2m\omega_n = 2\sqrt{mk}$.

When the resistance is zero, $r = 0$, then,

$$T = \frac{1}{\left|1 - \left(\dfrac{\omega}{\omega_n}\right)^2\right|} = \frac{1}{\left|1 - \left(\dfrac{f}{f_n}\right)^2\right|} \qquad (7.15)$$

As Figure 7.14 clearly shows, when $f/f_n < \sqrt{2}$ the force transmitted is greater than the applied force. At the point where the driving frequency and

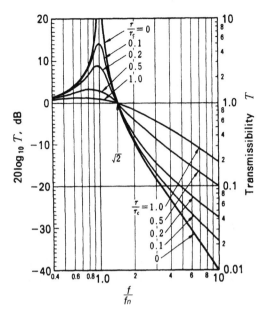

Figure 7.14 Vibration transmissibility vs normalised frequency.

the natural frequency are equal, the transmitted force theoretically becomes infinite because $r = 0$. Of course, damping resistance would prevent this situation arising, though it reduces the isolation efficiency to some extent, but it illustrates the problems associated with using springs as resilient mounts. When $f/f_n = \sqrt{2}$ then the transmitted force equals the applied force. At higher frequencies we start to get isolation, and for vibration isolation $f/f_n > 4$ should be the aim. The above discussion can be applied in the reverse direction in Figure 7.13, where vibration is transmitted from the floor to the mass m. Thus, in order to produce a quiet room that is isolated from vibration and from the transmission of structure-borne sound, the room considered as a mass m should be supported on a suitable spring system, which is generally called a **floating structure**.

B. *Vibration isolation design*

When a body whose weight is W (kg) is loaded on an elastic spring whose elastic modulus is k (kg cm^{-1}), the deflection δ is

$$\delta = \frac{W}{k} \text{(cm)} \tag{7.16}$$

The natural frequency of this vibration system, ignoring resistance, is

$$f_n = \frac{1}{2\pi} \sqrt{\frac{kg}{W}}$$

where g is the acceleration due to gravity, and from Equation (7.16)

$$f_n = \frac{1}{2\pi} \sqrt{\frac{g}{\delta}} \approx \frac{4.98}{\sqrt{\delta}} \text{ where } g \approx 980 \text{ (cm s}^{-2}) \tag{7.17}$$

Thus, the natural frequency f_n is determined by the static deflection δ. This relationship is shown in Figure 7.15.

The design procedure is as follows:

1 Knowing the frequency f of the vibration source, f_n is determined from Figure 7.14 in order to secure sufficient attenuation.
2 To realise this, the necessary deflection δ is obtained from Figure 7.15.
3 Select the vibration isolation material to maintain the δ while supporting the total weight W.

Since Figure 7.15 is based on the assumption that the material is perfectly elastic, it must be noted that except for steel springs, where Figure 7.15 has direct application, all other materials have a non-linear relationship between load and deflection so that dynamic deflections are smaller than static deflections. Therefore, the value of δ in Figure 7.15 must be increased by a correction factor indicated in Table 7.1.

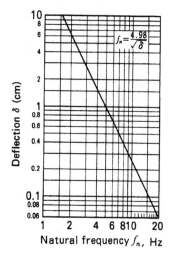

Figure 7.15 Static deflection vs natural frequency.

Table 7.1 Various vibration isolating materials

	Coil spring	Rubber isolator	Cork	Felt
Static deflection limit	Design free	Up to 10% of max. thick	Up to 6% thick (max. 10 cm)	–
Correction factor for static deflection	1	1.1–1.6	1.8–5	9–17
Effective frequency range	5 Hz or less	5 Hz or more	40 Hz or more	100 Hz or more
Allowable load (kg cm^{-2})	Design free	2–6	2.5–4	0.2–1.5

C. Vibration control materials

The materials listed in Table 7.1 are briefly described below.

a. Coiled steel spring

Usually the design properties of steel springs are clearly stated and a sufficiently large deflection can be obtained, therefore, they are indispensable for the isolation of low frequency vibration. A disadvantage, however, is their lack of damping, therefore a resistive element, such as an oil damper, must be introduced to suppress the amplitude at resonance. Also, longitudinal vibrations of the spring coil itself may occur with a resonance frequency given by

$$f_s = \frac{n}{2}\sqrt{\frac{k}{m_s}} \quad (n=1,2,3,\ldots) \tag{7.18}$$

where k is the elastic modulus and m_s is the moving mass of the spring. This may negate the effect of vibration damping and is called **surging**. As for the transmission of structure-borne sound between the spring and concrete floor, vibration isolation may not be expected due to the small difference in characteristic impedance (Table 1.1) as explained in Figure 1.8. However, this can be overcome by using a rubber isolator in combination with the spring.

b. Rubber isolator

This type has become popular in recent years. The reasons are:

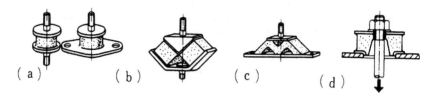

Figure 7.16 Examples of various rubber isolators.

1   Various reliable installation devices have been so developed that the selection of the appropriate isolator for the required use has become quite simple, as shown in Figure 7.16.
2   Much design data are available and there is a wide choice of elastic moduli to choose from.
3   Rubber has intrinsic damping properties due to internal friction, so that the resonance amplitude should not be excessive.

Although natural rubber has poor resistance to weather, oil and chemicals, synthetic rubber products are now available with a variety of characteristics, so that a type appropriate to the particular requirement can be selected.

c. Cork

The material quality may be neither constant nor homogeneous, therefore, it is difficult to carry out a precise design of isolation. Also, since cork cannot handle large deflections, the natural frequency may be limited to frequencies greater than or equal to 10 Hz, thus, instead of vibration control, cork may be more useful for preventing the transmission of structure-borne sound.

d. Felt

This material, except in lightweight systems, may not be used. The quality is so varied that comprehensive data are not available. Instead of vibration control, felt can be used to prevent the transmission of structure-borne sound, using its properties as a good packing material with low friction to maintain airtightness.

e. Air spring

These are often used in vehicles. Although the structure is complicated, their performance is very good. Air springs can be used as advanced vibration damping support for machinery and in special testing laboratories.

7.5   Vibration control of equipment and mechanical systems

A. *Vibration isolation of a machinery base*

Vibration control in the installation of equipment and machinery in build-ings is achieved by placing them together with the driving motor on a common bed, as shown in Figure 7.17, in such a way that the bed is two to three times heavier than the total weight of machinery, in order to reduce the vibration amplitude and to satisfy the condition of 1 d.f. Referring to Figure 7.15 for the total weight including the common bed, one can select a suitable vibration damping material or mechanism to produce the required deflection. It is recommended that δ should be 2.5–7.5 cm, using a system of a steel coil spring with a rubber pad, so that $f_n < 3$ Hz will be satisfied. If such machinery is to be installed on a high storey floor, a proper structural system should be designed to take the concentrated load of the machinery system directly on the floor beams, as shown in Figure 7.17.

B. *Vibration isolation in plumbing*

When equipment or machinery is loaded on such a vibration isolation base, as shown in Figure 7.17, the amplitude of the machinery itself generally becomes large, therefore, all terminals of the machinery connected to ducts, plumbing and wiring, etc. should be isolated from the vibration by inserting rubber pipes, canvas ducts and other soft and flexible materials to avoid rigid connections.

Apart from the machinery itself, noise and vibration are also generated by flowing fluid (air, water or their mixture) in pipes. Many portions of a plumbing system, such as right-angled bends, sectional changes, volume-control dampers, valves, water-flushing devices and other taps, air inlets

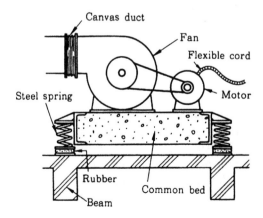

Figure 7.17 Example of vibration-isolation foundation for machinery.

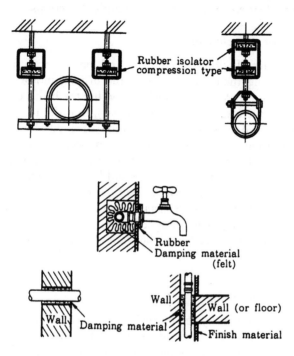

Figure 7.18 Examples of vibration isolation in plumbing.

and outlets, create resistance to flow and cause eddies, which always produce noise and vibration. Also, an abrupt change of pressure in a water
pipe can generate the well known **water hammer** effect. Of course, these
noises should be minimised but it is difficult to eliminate them completely,
so all plumbing fixtures should be isolated from the building structure with
resilient mounts. Some examples are shown in Figure 7.18.

7.6 Floating construction

A. *Floating floor and ceiling*

In order not to transmit the structure-borne noises caused by footsteps or
from other sources, a discontinuous floating construction is recommended,
as shown in Figures 7.19 and 7.20. Although dense glass fibre mats are used
as a resilient layer, rubber cushions may be more effective.

Figure 7.20 shows that the ceiling is also floated on rubber springs, but
it should be noted that there may be no advantage unless the floor itself is
floated.

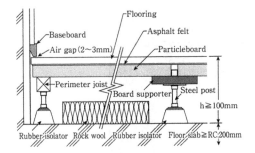

Figure 7.19 Floating floor for timber construction.

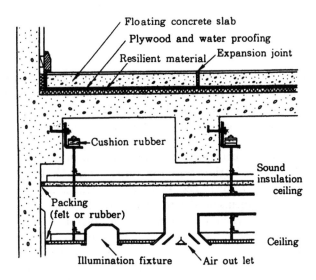

Figure 7.20 Floating construction for concrete.

B. Complete floating construction

When a high level of sound insulation is required, such as required for an acoustics laboratory, broadcasting or recording studio, it is essential that not only the door and ceiling but also the room itself should be completely discontinuous and floated from other parts of the construction, as shown in Figure 7.21.

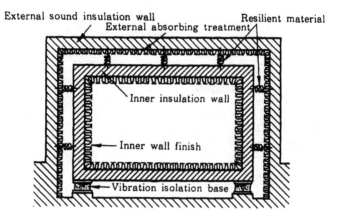

Figure 7.21 Complete floating construction concept. It is possible to omit the resilient material connecting both inner and external insulation walls.

PROBLEMS 7

1 When a wide, rigid wall vibrates in a piston-like manner, show that the velocity level (relative to $5 \times 10^{-8}\,\mathrm{m\,s^{-1}}$) of the wall equals the intensity level of the plane sound wave radiated from the wall.

2 List the devices effective in reducing structure-borne sound in a residential building.

3 Determine the single number rating of impact sound insulation of a floor using the ISO 717-2 method with measured data; 125 Hz: 62 dB; 250 Hz: 63 dB; 500 Hz: 57 dB; 1 kHz: 45 dB; and 2 kHz: 27 dB.

4 Explain the basic requirements needed to prevent vibration or structure-borne noise transmission when using rubber as an isolator.

5 Design a vibration damping foundation for a ventilation fan that is 110 kg in weight, rotating at 617 rpm and driven by an electric motor of 120 kg weight, rotating at 1800 rpm. What is the weight needed for the common bed and what should be the deflection of the resilient support?

8 Noise and vibration control in the environment

In daily life outdoors, we encounter noise and vibration generated by, for example, aircraft, road and rail traffic, and indoor noise and vibration sources such as mechanical ventilation systems, elevators and other items of equipment. It is against all such sources that countermeasures are sought. Even in residential accommodation there are many kinds of electrical appliances: dishwashers, food processors, washing machines, which are on the increase and give rise to annoyance.

In principle there is no different approach for noise reduction outdoors and indoors. Therefore, we shall discuss procedures for noise and vibration control that are common to both environments.

8.1 Basic strategy

In solving environmental noise problems we need to consider the following: (1) the sources of noise and vibration; (2) the propagation paths; and (3) the input to the receivers. Countermeasures should be taken to deal with each of the above in a step-by-step logical sequence.

Suppose a fraction of the energy W generated by a noise or vibration source produces a sound level L_0 (dB) at a receiving point, depending on the acoustic nature of the spaces in which source and receiver exist. Assume that the sound level is reduced by R (dB) along the propagation path, as described in Chapters 5, 6 and 7, then, we have

$$L_W - R = L_0 \text{(dB)} \tag{8.1}$$

where L_W is the power level of the source, as Equation (1.17b).

On the other hand, at the receiving point, if L (dB) is the level necessary to satisfy this required environmental condition and $L_0 < L$, then there is no problem, but if $L_0 > L$ complaints might be expected. Therefore, the required noise reduction R_r, is given by

$$L_0 - L = R_r \text{ (dB)}$$

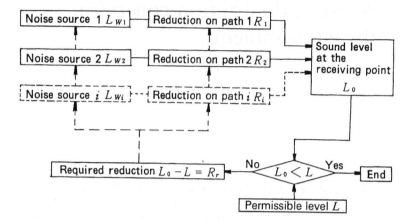

Figure 8.1 Flow chart of noise and vibration control.

When there are many sources, then the energy from each source must be added together at the receiving point as shown in Figure 8.1.

When considering a transmission path in which barriers or attenuators are in series, each with their individual reduction R_i, so that $R_r = \sum R_i$ is the total reduction in the path, then it is only necessary to add together sufficient reducing elements to meet the requirement. When there are many sources in parallel, after calculating the attenuation in each path, the value of L_0 is again obtained. This is a basic procedure for noise reduction by the **energy flow method**, described in Figure 8.1 as a flow chart.

It should be noted that in the process of calculating dB values, it is not necessary to show the sound pressure level at individual points but to show the acoustic energy ratios. Therefore, even though a complicated phenomenon peculiar to wave motion, such as interference due to reflected sound, may occur locally, it is of no concern and only the energy decay is relevant. As discussed in Chapters 5, 6 and 7, energy attenuation can be analysed and controlled by using wave acoustics. With fundamental knowledge properly applied, however, effective noise control measures can be executed using simple geometrical acoustic calculations. This is because noise is seldom a pure tone in nature but in most cases has a continuous spectrum with random amplitudes and phases, although, when dealing with pure tone noise, its reduction needs to take account of wave acoustics.

8.2 Determination of required reduction

In this section we discuss the procedure for determining the required reduction R_r to be added to the existing reduction R, which is calculated or measured. The value of L_W in Equation (8.1) needs to be precisely defined and the permissible level L should be suitably selected. This procedure plays an important role in the assessment of noise pollution and in the estimation of the quality and magnitude of noise control in a new project.

A. Locating noise and vibration sources

a. Source extraction

Where there are several noise sources, which together give rise to a problem, a decision has to be made as to which is the most serious and action taken to reduce it.

During the process of measurement and analysis of the noise, the various sources must be switched off, one at a time, until the principal source is found, against which control measures are to be taken. The other sources should then be investigated in a similar manner, always reducing the noise from the most prominent source. If it is not possible to carry out actual acoustic measurements, then judgment must be used with the aid of existing data as to which source is the largest contributor to the noise problem, otherwise any sources which are missed can give rise to trouble later on.

b. Noise source quantification

If the power level in every octave band along with the directivity of every sound source is obtained, one can proceed to plan and design to reduce the noise. If, after acquiring the relevant published data, they are still insufficient, actual measurements are highly desirable.

When the sound source is pinpointed by means of measurement and analysis following the procedure described above, the frequency characteristics of the equivalent power level of the source can be found.

Generally the required frequency characteristics are presented as octave-band levels. If the values are obtained from other forms of analysis, they should be converted to octave-band widths (see Section 2.1.B).

[**Ex. 8.1**] A basic noise survey was carried out in a large factory, in order to plan the noise control strategy. The measured noise levels in a frequency band, when the operating machines $M_1, M_2, M_3, \ldots$, were stopped one at a time, are as follows:

Machine stopped	None	M_1	M_2	M_3
Band level (dB)	100	93	99.5	99.7

From this data, the source producing the maximum noise level is seen to be M_1, which, therefore, needs to be reduced. Further, after stopping the machines one after another, the following results were obtained.

Machine stopped	M_1	$M_1 M_2$	M_1, M_2, M_3
Band level (dB)	93	90	86

From the above, the noise level produced by each one of the three machines can be estimated as follows:

Machine	M_1	M_2	M_3	All other sources together
Band level (dB)	99	90	88	86

B. Locating the transmission path (calculation of existing attenuation)

It is now necessary to understand the nature of the noise transmission path and to calculate the existing reduction R in Equation (8.1) along the path. When there is only one path, for example through an air-conditioning duct, the natural attenuation can be calculated precisely. Where paths exist in parallel, the reduction in each path should be calculated and summed at the receiving point.

At the actual point where the noise is of concern, the reduction R can be obtained from the difference of the measured values between two points, one close to the sound source and the other at the receiving point. However, it is generally difficult to determine the sound paths, e.g. airborne or structure-borne. Also, there are much more complicated situations as in the following example, which requires special experimental approaches or techniques. Otherwise, the appropriate noise-reduction procedure cannot be determined.

[Ex. 8.2] In a conference room within a municipal office building, the noise from the plant room adjacent to the conference room became intrusive. In this case the noise transmission can be associated with the following paths, (a) airborne noise transmitted through the partition wall; (b) air flow in the duct; and (c) structure-borne noise from the plant foundations. Hence, measurements of sound pressure levels were obtained, derived from spectrum analysis, as shown in Table 8.1.

Table 8.1 Experimental results for separating the airborne and the structure-borne noise

Experiment	Sound pressure level (dB)		
	Plant room	Conference room	Difference
(1) Noise analysis in both rooms during normal plant operation	87	62	25 dB
(2) While plant stopped, measurements in both rooms with white noise generated by a loudspeaker in the plant room	105	46	59 dB

From the results, (a) it is clear that the airborne sound insulation is sufficient. As regards (b) and (c), after proper attenuation in the duct was determined by calculation, and localisation of the noise source in the conference room checked by ear and confirmation that the air outlets from the ducts were not the noise sources, the main source of noise was judged to be due to (c) because of inadequate vibration damping of the air-conditioning plant.

C. Determination of permissible noise level

The permissible values at the receiving point can be determined by the evaluation method discussed in Section 2.2. NCB-curves (Figure 2.6, Table 2.5) for steady noises, and Figure 2.12 and Table 2.7 for vibration, should be applied.

When there are two or more noise sources or transmission paths, each individual permissible level should be determined, so that the total value will satisfy the criteria as outlined in the flow diagram of Figure 8.1.

For example, in a situation where noise from several sources, i.e. (1) outdoor noise; (2) air-conditioning noise; and (3) adjacent-room noise are transmitted then, if they are all meaningless noises, the ratio of noise energy from each path can be determined in the usual way depending on the ease of control. If, however, an intermittent noise is included, its peak value must be used, or if it is an impact noise with a large statistical fluctuation, it should be handled separately. When meaningful noises, such as speech or music are included, it is desirable that each energy contribution is 5–10 dB lower than the other meaningless noise energy.

D. Calculation of required reduction

As shown in Figure 8.1, the additional reduction required, R_r, in each path should be calculated in each octave band in the appropriate frequency range according to the methods described above.

8.3   Organisation of noise and vibration control

In order to realise the required reduction obtained with the aid of the above procedures, the measures should be planned according to the flow chart shown in Figure 8.2. If the sequence is followed in the wrong order, then all efforts may result in failure and may incur more expense and waste time.

A. Dealing with noise and vibration sources

Noise and vibration problems never occur if sources do not exist. Generation of noise and vibration should be eliminated wherever possible at source when selecting equipment and machinery for installation in a building or

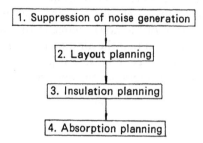

Figure 8.2 Sequential order for planning of noise and vibration control.

factory. When equipping a building or factory, those versions of equipment that generate unacceptable levels of noise and vibration should be avoided by referring to the manufacturers' data. Machine specification, and noise and vibration criteria should be determined from the standpoint of environmental planning.

Also, noise and vibration nuisance produced by traffic and industry should be controlled by proper land-use zoning and road planning as an important strategy of city and regional planning. Figure 8.3 shows a good sample of car noise control in a residential area by means of planning layout (see Lit. A6).

B. *Planning layout*

All noises are reduced, more or less, with distance. Therefore, city/urban and architectural planning may have, as their basis, zoning and site selection

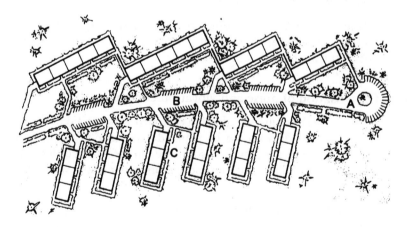

Figure 8.3 Layout of a residential street for vehicle-noise control. A: end loop of the street; B: external car parking areas; and C: pedestrian access to the buildings (Doelle 1972).

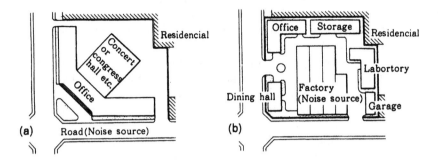

Figure 8.4 Examples of design layout for noise control: (a) preventing street noise for quiet facilities; and (b) preventing factory noise nuisance.

associated with effective layout of roads, buildings and rooms with noise reduction as a prime aim. In principle, zones and those rooms which need to be quiet, should be located as far as possible from any offending noise sources. This also applies to indoor sources such as mechanical services.

As shown in Figure 8.4, it is advantageous to locate buildings and rooms which are not sensitive to noise, such as a storeroom or workshop, so that they will act as noise barriers in the transmission path from the source, so that natural reduction can be greatly increased. If basic planning is given sufficient thought, R in Equation (8.1) may be sufficiently large to satisfy the required criteria without any other measures taken. In contrast, if the layout does not reflect any acoustic planning, complaints may occur at a later date and, in fact, there have been many cases where satisfaction has not been achieved, even with vast expenditure, in an attempt to correct previous planning errors.

C. Noise insulation planning

After reducing the noise at source as much as possible, and transmission paths to the receiver have been identified, effective sound insulation by means of, for example, noise barriers, silencers and vibration isolation devices should be planned using the methods described in Chapters 5, 6 and 7.

As discussed before, when identifying the transmission path, serious consideration must be given to the following:

1   In the open air, when the diffraction attenuation by a barrier is to be calculated, it is necessary to make sure that there is no other reflection path at the same time.
2   When calculating the transmission loss of an exterior wall, any opening such as that provided for ventilation must be taken into account.

3   In the case of partitioning, whether or not it extends to the soffit of the floor above must be checked. Also, any transmission path, due to a ventilation or air-conditioning duct above a suspended ceiling, should be investigated.

4   The insulating capability of doors opening onto a corridor, which represents a sound transmission path, should be considered. Particularly, air vents in the form of louvres should be examined.

5   Since any openings offer little in the way of sound insulation compared with a wall, their location must be considered very carefully and so, for example, a corridor or anteroom should be treated with sound absorbents to create a sound lock, if necessary.

D. Sound absorption treatment

In the spaces, such as those of the source room and receiving room, or the corridor which is a transmission path, if the sound energy is absorbed by applying absorptive treatment, the noise level L_0 in Equation (8.1) is reduced according to Equations (6.2)–(6.4) in Chapter 6. Then, the amount of sound insulation required is less. Moreover, in noisy factories, the subjective noisiness experienced by a worker is much reduced, due to the more rapid decay of reflected and diffused sounds, though the fact is related only to the small difference in the calculated or measured noise level.

The effect of absorptive treatment depends upon the amount of absorption present before treatment, since the noise level is reduced by $10 \log_{10} (A_1/A_0)$ (dB), where only the term related to absorption needs to be considered in Equation (6.3), where A_0 and A_1 are the total absorption in the room before and after treatment, respectively.

[Ex. 8.3] In a machine room where the average absorption coefficient $a_0 = 0.05$ is changed to $a_1 = 0.3$ by the absorptive treatment, the effect is a noise reduction of $10 \log_{10} (A_1/A_0) = 10 \log 6 = 8$ dB. If more absorption is added up to $a_2 = 0.5$, the effect is increased by only $10 \log_{10} (0.5/0.3) = 2$ dB.

It should also be noted that the cost required for the latter, 0.3 to 0.5, is generally more than the step from 0.05 to 0.3.

E. Electro-acoustic devices

a. Active noise control (ANC)

By using a microphone and loudspeaker system, noise can be cancelled by generating sound in antiphase, a method which is called active noise control. Although this idea had been considered nearly 100 years ago, it is now possible using digital technology. It is being used effectively for localised spaces such as air-conditioning ducts, and is progressing to wider applications, e.g. noise barriers in open air, etc., however, its reliability, durability and the maintenance cost of equipment, etc. need to be carefully assessed.

b. Masking noise generator

Meaningful noises, such as speech, can be masked by meaningless noise, which has a continuous spectrum like white noise. Electro-acoustic equipment can be used to generate noise in a room, where the masking of undesired speech from a neighbour in an office or a hotel is needed. Air-conditioning noise is effective for this purpose, so that determination of an appropriate noise level (see Section 8.2C) needs to be taken into account.

c. Background music

Background music has been proposed for masking noise and reducing the annoyance caused by it. Although this idea may work to some extent, when the noise level becomes too high, the background music itself becomes annoying. Of course, **environmental music** can be used for purposes other than noise control to proper effect, so background music itself may not be considered an adequate means of noise control in the environment.

8.4 Examples of noise control planning

A. Outdoor noise control

Taking as an example, a hotel intended for a site alongside a busy road, which has, in addition, trams as well as road traffic, as shown in Figure 8.5.

Actual measurement of noise was carried out and the required reduction for noise control investigated, with results shown in Table 8.2 and Figure 8.6. Since the required permissible level is specified as NCB-30, the required reduction is 45 dB at 500 Hz. This is found from the left-hand side of Equation (6.2); therefore, the composite transmission loss of the facade is given by the first term on the right-hand side of the equation.

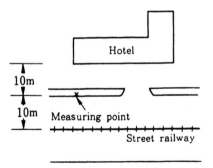

Figure 8.5 Situation of a hotel building with regard to control of traffic noise (see Table 8.2 and Figure 8.6).

Table 8.2 Example of outdoor noise prevention calculation (Figures 8.5 and 8.6)

Octave band mid-frequency	63	125	250	500	1k	2k	4k (Hz)
(1) max. value from noise measurement analysis (dB)	89.0	84.5	79.5	83.0	74.5	69.5	65.5
(2) (1) − 3 dB = noise level at outside surface of wall (dB)	86.0	81.5	76.5	80.0	71.5	66.5	62.5
(3) permissible noise level in rooms (dB)	55.0	46.5	40.0	35.0	32.0	38.0	25.0
(4) (2) − (3) = required reduction (dB)	31.0	35.0	36.5	45.0	39.5	38.5	37.5

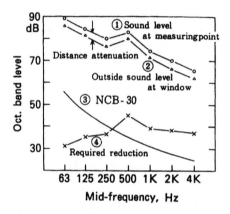

Figure 8.6 Calculation of traffic noise control (see Figure 8.5 and Table 8.2).

B. *Factory noise control by means of the building envelope*

Since the noise source is inside the building, the exterior wall has to provide sufficient insulation against transmitted noise in order not to cause any annoyance to neighbours. In Figure 8.7, a glass block window in a heavy concrete wall is shown located at 2 m from the site boundary. Assuming a permissible level of NCB-40, the required attenuation calculation is shown in Figure 8.8.

In this example, the source is a diesel engine whose total power is 3365 hp, and the noise power is assumed to be proportional to the hp value obtained from measurement on a similar engine. The power-level curve is labelled ① in the figure. Using Equation (6.7), the room total absorption $10 \log_{10} A_1$ and TL of the glass block can be deduced, then curve ③ is the resultant noise power level radiated externally from the window. Attenuation due to distance to the site boundary, is found to be −6.5 dB from Equation (5.11) and Figures 5.4–5.5, thus curve ④ is obtained. Referring

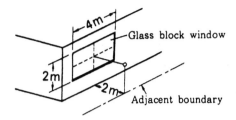

Figure 8.7 Geometry of noise-generating window in a factory.

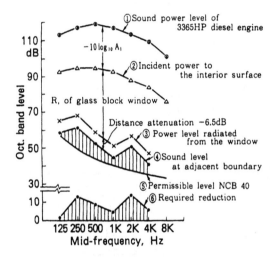

Figure 8.8 Calculation of noise-nuisance control for a factory (see Figure 8.7).

to the permissible level ⑤ the required attenuation ⑥ is obtained. A fixed-glass window, in addition to the glass block, resulting in a double window configuration, outlined in Section 6.6, may satisfy the requirements. When a room, as in the above example, is made airtight, it is inevitable that air-conditioning will be installed; therefore, attenuation of noise transmitted though air ducts must also be carefully controlled, and this will be described later.

C. Noise control of indoor noise sources

After control of the source noise, as discussed above, acoustic energy still remaining or generated indoors should be dealt with by enclosing the equipment using a cover, hood or barrier. Figure 8.9 shows examples, which are effective for noise generated during machine operation in factories. While the materials used for them should have adequate TL, it is

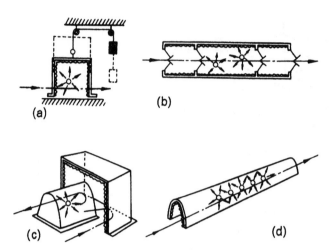

Figure 8.9 Barriers and enclosures for noise-generating sources; (a) closed and movable arrangement; (b) anteroom arrangement; (c) directional absorbing
arrangement; and (d) absorbing tunnel.

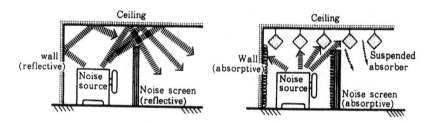

Figure 8.10 Noise screen and sound absorption treatment in a room: (a) before
treatment; and (b) after treatment.

absolutely necessary that they are lined with an absorbent, as shown in
Figure 8.10. It is also important that isolating partitions divide the shop
floor into smaller sections in which there is ample absorption, in order to
improve the workers' auditory environment, in the sense that he or she
is more likely to be annoyed by noise other than that from his/her own
machine.

D. Application of scale model and computer model studies

The most essential task in the flow diagram of noise control planning,
shown in Figure 8.1, is the finding the reduction R in the transmission paths.
In the above examples, A and B, it is easy to find R, but in case C it is rather
difficult to predict R by calculation. Also, in the open air where there are

very complicated configurations, when barriers are in a complicated terrain, and when the source is distributed such as in the case of a highway, or is moving, e.g. a train or aircraft, it is necessary to carry out the analysis using the methods described in Chapters 5, 6 and 7. The analysis may need to use a scale model or a mathematical model closely approximating the actual situation, the former for acoustic measurements and the latter for computer simulation. This can be used to assess noise propagation and attenuation.

Figure 5.13 in Chapter 5 shows an example of a computer simulation. Figures 8.11 and 8.12 are examples of a comparison between the results obtained by computer and acoustic models, the latter is shown in Figure 8.13.

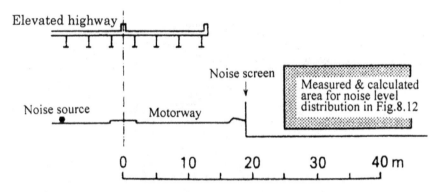

Figure 8.11 Cross-section of an elevated highway showing the area of model measurement and computer simulation for noise propagation.

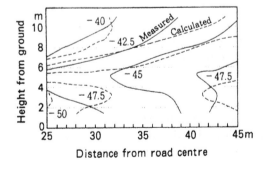

Figure 8.12 Comparison of results between model measurements (solid lines) and computer simulation (dotted lines) for noise propagation from the highway shown in Figure 8.11. Levels show relative (dB) values referred to the level at the point 1m from the source (Fukushima *et al.* 1990).

Figure 8.13 Scale model for acoustic measurement of the elevated highway shown in Figure 8.11.

E. Noise control of air-conditioning ducts

a. Noise generated at the air grill

Both air outlets and inlets generate noise in the same way, independent of air flow direction, by impinging air flow on the grill or diffusers. The power level per 1 m² of the duct neck section is empirically proportional to about the sixth power of the wind speed v (m s^{-1}) at the louvre surface. This can be expressed mathematically as follows

$$L_W \cong 30 + 60 \log_{10} v (\text{dB m}^{-2}) \tag{8.2}$$

This noise can be reduced only by decreasing the wind speed. Therefore, calculating the sound level in the room using Equations (6.3) and (3.50), the wind speed should be decreased so that the generated noise level is lower than the permissible value.

b. Air flow noise in ducts

At bends, junctions and dampers, eddy currents occur and generate noise, whose level is also proportional to the fifth or sixth power of the air speed. Although it, too, can be reduced by lowering the speed, since the demand for high-speed ducting is common in current practice, the noise reduction of air flow is a significant problem. In a building that requires more strict noise control, low-speed ducting (wind speed 5–10 m s^{-1}) should be adopted with the air flow as smooth as possible and with a duct wall which, with careful design and construction, does not vibrate at low frequencies.

c. Noise transmitted through duct wall

The duct wall should have adequate TL depending upon the duct location, so that noise may neither be transmitted into the duct nor emitted externally. Special attention is needed when using glass fibre ducting, which has extremely low transmission loss.

d. Cross talk between rooms through common duct

When there are outlets in nearby rooms in a common duct, sounds can be heard from one room to the other via the duct. To avoid this cross talk, the distance between the outlets should be kept sufficiently large by means of a tortuous route or separate ducts should be provided. Silencers or mufflers should also be installed between the outlets.

e. Fan noise

This is the main source of noise in low-velocity ducting. The power level of the generated noise is proportional to the nominal power K_w of the driving motor as follows,

$$L_W = 89 + 10 \log_{10} K_w + 10 \log_{10} \left(\frac{P}{25} \right) \text{ (dB)} \tag{8.3}$$

where P is the static pressure (mmAq). Although there is another formula that takes into consideration the type of vane and flow volume of the fan (see Lit. A7), Equation (8.3) gives an approximation sufficient for this purpose. The relative values of the octave-band level can be estimated from Figure 8.14. After calculating the natural reduction of this noise level by the duct system, silencers or mufflers may be added if the reduction is insufficient.

The procedures follow, with the illustrated example shown in Figure 8.15.

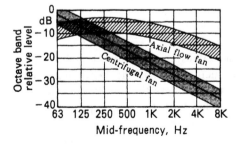

Figure 8.14 Frequency characteristics of generated noise power from fans in a duct.

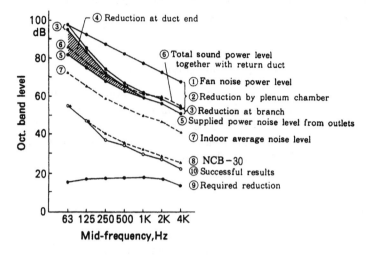

Figure 8.15 Determination of noise control in the air-conditioning system shown in Figure 8.16.

① It is recommended that the noise level generated by a centrifugal fan is obtained using data supplied by the manufacturer. When there is no data, Equation (8.3) and Figure 8.14 can be used for estimation.

② The reduction required for the duct system is shown in Figure 8.16, when it has only an absorbing chamber. The power level at a point just before the branch ⓐ is obtained from Equation (6.47).

③ The reduction at the branch is calculated by means of the section area ratio of the ducts. 2.4 dB is obtained at point ⓐ.

④ The reduction at the open end, where the noise power is radiated from 4 grills ⓖ into the room, is obtained from Figure 6.27.

⑤ The sound power level from the outlets was obtained by the above calculation.

⑥ Adding on the noise power ⓡ radiated from the return grills, calculated by the same method described above, the total sound power level supplied to the room can be determined.

⑦ The average noise level in the room can be obtained by Equation (6.3).

⑧ Shows the permissible level NCB-30.

⑨ The difference between the values of ⑦ and the permissible level NCB-30 is the required reduction. Then, the bend lining for high frequencies, and resonator-type mufflers for low frequencies, such as shown in Figure 6.29, are added with the aid of proper design (see Section 6.9.B–F), as shown in Figure 8.16.

⑩ Shows successful results of this design procedure, where an average noise level is produced by the air-conditioning system in the room.

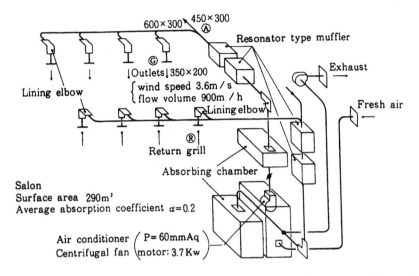

Figure 8.16 Example of air-conditioning system in which noise control is to be determined by calculation in Figure 8.15 (Hayakusa 1990).

When a silencer or muffler is to be added, though it is rather easy at high frequencies, it is very difficult to find sufficient space for a silencer capable of attenuating the low frequencies, unless thought has been given to the problem right at the planning stage.

If the noise level at a point near the duct opening is of concern, the direct sound power from the opening should be added in, using the calculation of attenuation with distance given by Equation (3.44).

f. Structure-borne noise transmission by ducts

In addition to the airborne noise carried by the duct, the duct wall transmits structure-borne noise and, furthermore, the duct-supports transmit noise to the building structure. Therefore, even if silencers or mufflers are installed, it is said that only a maximum of 25 dB for frequencies below 100 Hz, and 50 dB for 1000 Hz or above may be attenuated by a continuous duct made from steel plate, provided that vibration damping is installed, as shown in Figure 7.18. Further noise control, such as canvas ducts and concrete ducts, can be inserted so that vibration and structure-borne sound can be isolated. It is most important that thought is given to vibration damping and structure-borne isolation in duct design in order to obtain a very quiet room.

PROBLEMS 8

1   Give your opinion on noise nuisance as an environmental problem.

2   Give all possible means of reducing road traffic noise.

3   Describe your ideas on methods of avoiding noise nuisance from aircraft.

4   List the methods of controlling noise generated indoors, applying the concept that the greatest noise has greatest priority.

5   In an apartment, there are complaints about the neighbour's TV, which produces a measured sound level of 45 dB in the 500 Hz octave band. There is also air-conditioning noise of 42 dB in the same frequency band. How much reduction must be applied to each of the two sources to meet NCB-35 and eliminate any meaningful sound from the TV?

6   A very noisy factory measuring $20 \times 30 \times 5 \, m^3$ has a measured noise level of 94 dB and reverberation time of 5.8 s in the 1 kHz octave band. How much absorptive treatment is needed to reduce the noise level to 86 dB in that band?

7   In the above factory, what control measures should be considered before introducing absorptive treatment? List and explain, stating the order of priorities.

8   List the sources of noise to be controlled in the air-conditioning system and explain how the control is achieved.

9   In Figure 8.16, how does the noise level differ from the calculated value shown in Figure 8.15, with direct sound from an outlet grill at a point 1 m from it and at a point 2 m from the grill?

9 Acoustic design of rooms

Although any room may have an optimum acoustic environment, depending on the purpose for which it is intended, it is difficult to determine the ultimate goal because the final evaluation still relies on a subjective auditory sensation. However, with the knowledge so far gained, one can design the space acoustic without any serious defects. In this chapter we discuss the further research necessary to create a better acoustic environment in a room.

9.1 Design target

General requirements for the acoustic design of a room are as follows:

1 any intrusive noise should be avoided;
2 speech intelligibility should be satisfactory;
3 music should sound pleasing and have warmth;
4 a uniform distribution of sound should be observed throughout the whole room;
5 there should be no defects such as echoes or flutter.

When all the above conditions are satisfied, the design can be considered successful.

First, noise should be completely controlled. This is a prerequisite necessary in room design. Although there are much data available for speech intelligibility, as discussed in Chapter 1 (see Section 1.14), the question arises as to the possibility of providing sonorous music at the same time. Further research on which physical condition can produce the best music performance, in addition to the reverberation, as described in Chapter 3, is needed (see Section 3.6).

On the other hand, current knowledge and techniques can improve the sound field distribution, avoiding undesirable phenomena such as echoes, foci, etc. The remaining problem is how to integrate the acoustic and visual treatment of the room into something which is aesthetically totally satisfying.

9.2   Design of room shape

A. Fundamental principles

The shape of the room is the most fundamental factor in influencing its acoustics. So a building requiring good acoustics needs to be examined at the basic planning stage, with advice from a good acoustic consultant. As discussed fully in Chapter 3, depending on room size, wave acoustics should be used to assist in the design of small rooms and geometrical acoustics for larger spaces.

a. Dimensional ratio for a rectangular room

In a small rectangular room, the natural frequencies given by Equation (3.13) should not degenerate and should be uniformly distributed. A simple integer ratio of the three edge-lengths must be avoided, as shown in Figure 3.5. The golden rule, $(\sqrt{5} - 1):2:(\sqrt{5} + 1)$, or its approximate ratio such as 2:3:5, has been recommended over many years. Generally it is considered appropriate to select values of $2^{n/3}$ or $5^{n/3}$. For example, $1:5^{1/3}:5^{2/3} \approx 1:1.7:2.9$ is a handy ratio to apply in practice.

b. Surrounding wall shape and sound reflection

When sound is reflected from a wall surface that is dimensionally large compared to the wavelength, the reflected sound can be visualised in the same way as light reflected from a mirror as discussed in Chapter 1 (see Section 1.5.B). So, if the wall is concave, the reflected sounds are concentrated; on the other hand, if the wall is convex, they are diffused as shown in Figure 9.1. Therefore, in order to have good diffusion, concave surfaces that are large compared to the wavelengths should not be used. It is no exaggeration to say that almost all concave surfaces are, acoustic, potentially hazardous.

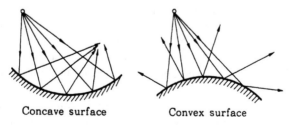

Concave surface   Convex surface

Figure 9.1 Reflection from a curved surface.

c. Prevention of echo and flutter

At a corner where two planes intercept orthogonally, either indoors or out-doors, sound waves are reflected back in the reverse direction to the incident one, so echoes can arise as shown in Figure 9.2. In order to avoid echoes completely, the surface shapes must be created so as to diffuse or absorb the incident sounds as shown in Figures 9.6 and 9.7.

As mentioned in Chapter 3 (see Section 3.5.B), a multiple reflection between two parallel planes may cause flutter echoes. Although it is easy to avoid them by making the two planes non-parallel or the surface undulated

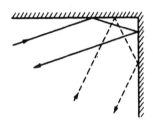

Figure 9.2 Reflection at a corner.

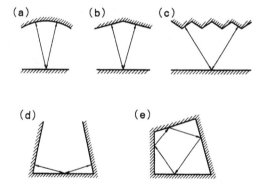

Figure 9.3 Shapes causing flutter echo.

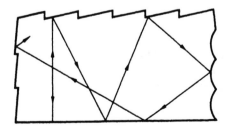

Figure 9.4 Shape causing no flutter echo.

with dimensions close to the wavelength of sound, the shapes shown in Figure 9.3 may cause flutter echoes when the wavelength is small compared with the pattern elements of the surfaces. Figure 9.4 shows a preferred shape. It is essential to diffuse sound waves in the frequency range above 1–2 kHz in order to prevent an audible flutter.

B. *Selection of local shapes*

a. *Profiles of sections*

1   *Ceiling*: Reflected sound from a ceiling plays a significant role in rein-forcing the direct sound, and is particularly important for listeners in rear seats. However, a concave domed ceiling makes the sound distribu-tion worse, as shown in Figure 9.5(a). This situation can be overcome with a combination of convex shapes, as shown in Figure 9.5(b). When the client insists on a domed ceiling, the radius of curvature should be at least twice as large as the ceiling height.

Since echoes should not occur where the ceiling meets the rear wall, as shown in Figure 9.2, a tilted wall, Figure 9.6(a), may still cause problems; therefore, a design such as the one shown in Figure 9.6(b) is recommended.

2   *Balcony*: When a balcony is provided in order to increase audience capacity, the acoustics under the balcony becomes generally worse because the direct sound is attenuated owing to long-distance propa-gation over the audience, and, hence, the sound pressure levels decay

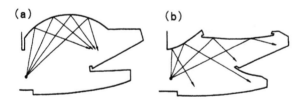

Figure 9.5 Longitudinal sections of auditoria: (a) poor ceiling; and (b) good ceiling.

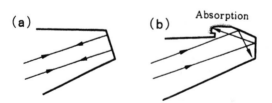

Figure 9.6 Rear wall corner design: (a) poor; and (b) good.

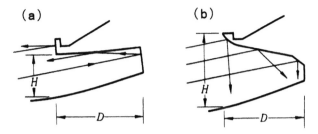

Figure 9.7 Sections of balconies: (a) poor; and (b) good.

owing to a lack of effective reflection from walls and ceilings. Moreover, the reverberation time is shorter due to the smaller volume per audience member and the weak diffused sounds. In order to improve this problem, the depth D of the balcony in Figure 9.7 should be as short as possible and less than twice the maximum height H (if possible, equal to the opening height H). It is also advisable to design the seating under the balcony so that a person sitting in any seat can see as much of the main ceiling as possible.

The balcony front should be designed so that any echo or focusing may not occur either in plan or in section, while the shape and material of the balcony soffit should be an effective reflecting surface.

3   *Floor*: Since a floor with seating has large absorption, direct sound is rapidly attenuated as it propagates over the absorbing surface because the porous-type absorption provided by chairs and audience is efficient at high frequencies. In addition, the spaces between rows of seating are found to cause resonance in the low-frequency range from 100 to 200 Hz, which produces remarkable attenuation. In order to overcome the situation, there is no other way but to increase the floor slope so that the direct sound is not interrupted by the front seats. The method of determining the floor slope is nothing more than a problem of having a sightline to the stage. One means of defining this is shown in Figure 9.8. In the figure, γ is the angle of elevation of the sound source above P_0, the person sitting in the front row. With the requirement that γ is a constant, the curve so generated, which starts at P_0 and continues to a receiver at P_n with distance d_n, is called a logarithmic spiral. Then, using polar coordinates (r, θ), by definition

$$r = r_0 \exp \theta \cot \gamma \approx r_0 \exp \frac{\theta}{\gamma}$$

$$\therefore \quad \theta = \gamma \ln \frac{r}{r_0} \approx \gamma \ln \frac{d}{d_0}$$

(9.1)

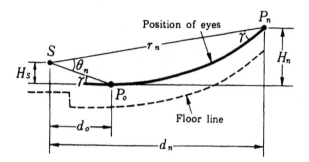

Figure 9.8 Seating floor section (Cremer Lit. B27).

Since $H_s \approx d_0 \gamma$, the height H of P from P_0 is obtained

$$H = d_0\gamma + d(\theta - \gamma)$$

$$= \gamma \left[d \cdot \ln \frac{d}{d_0} - (d - d_0) \right] \qquad (9.2)$$

where d_0 is the distance from the source to the origin P_0, the eye position. The condition where the angle γ is larger than 12–15° is said
to be desirable (Cremer and Müller Lit. B29). The floor slope, to be
equal to θ, becomes steeper as it progresses from the stage towards
the remote seating, and it is possible that this may conflict with local
building regulations.

b. Plan shape

As with the ceiling profile design, it is advisable to obtain the first reflected
sound from the wall nearest to the sound source, then gradually change
the wall finish towards a surface that diffuses or absorbs. Concave surfaces
should be avoided. Since an elliptical or circular plan, in particular, always
creates serious problems, if still required, a combination of convex elements
as shown in Figure 9.9(b) should be considered.

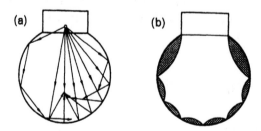

Figure 9.9 Treatment of a circular plan shape: (a) poor; and (b) good (Knudsen &
Harris, Lit. Al).

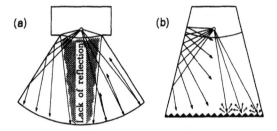

Figure 9.10 Fan shape plan: (a) poor; and (b) good.

A side-wall plan with a fan shape is often preferred by architects. However, it may produce too few lateral reflections, which play an important role in the subjective acoustic impression and may also give rise to echoes from the rear wall, as shown in Figure 9.10(a). Therefore, adequate diffusion and absorption should be considered, as shown in Figure 9.10(b). Generally, a space that is irregular or asymmetrical in plan is not easy to design, but is desirable from the point of view of sound diffusion and acoustic quality; therefore, the examples in Figures 9.22(b) and (c) should be noted.

c. Sound-reflecting panels

In multi-purpose halls, movable sound-reflecting walls and ceiling are often installed around the stage as an orchestra shell. By using these, the sound energy, which would otherwise be absorbed in the space at the rear of the stage, which has high absorption, is reflected back to the players, creating better ensemble, and providing improved listening conditions for the audience.

Although such reflecting panels should be as heavy as possible, a compound laminate-layer construction of plywood and damping rubber sheet has rather good reflection characteristics for low weight. The reflecting surface should consist of flat or convex units whose dimensions are comparable to the sound wavelengths (Figure 9.11). For example, at frequencies lower than 100 Hz, the width required is more than 3.4 m.

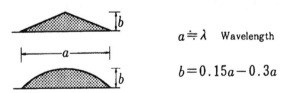

$a \doteqdot \lambda$ Wavelength

$b = 0.15a - 0.3a$

Figure 9.11 Dimensions of a diffusing element.

In large halls, in order to reduce the time delay of reflected sounds, many reflecting panels are often suspended from the ceiling over the stage and audience (Figures 9.23 and 9.25). Reflection characteristics of such devices should be studied (Maekawa & Sakurai 1968; Rindel 1991).

d. Diffusion surface or units

For the purpose of improving sound diffusion in a room, the wall or ceiling can be given a zigzag profile. Alternatively, cylindrical, spherical, pyramidal modelling or boxes of one sort or another, as well as various uneven irregular-shaped units, can be installed along the boundaries. In fact, any shape will work so long as the wavelength of interest is of the same order as the dimensions of the irregularity. Figure 9.11 may be a useful reference for designing such a unit. It is advisable to use various types and sizes so that a wide range of frequencies can be diffused. Figure 9.12 shows an example whose effectiveness has been measured by means of a scale model.

Schroeder (1979) proposed a special configuration developed from number theory, Figure 9.13(a), and showed that it reflects diffusely for normal incident sound, such as shown in Figure 9.13(b). This idea is currently being

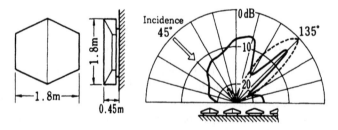

Figure 9.12 Examples of the reflection characteristics of a diffusing element measured by a scale model: (a) actual dimensions of diffusing element; and (b) directivity of reflected sound (actual frequency 250 Hz). Dotted line shows the case without any diffusing element (Maekawa *et al.* 1965).

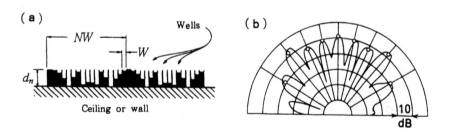

Figure 9.13 Example of diffuse reflection surface based on quadratic residues (Schroeder 1979): (a) section of the surface; and (b) directivity of reflection when the wavelength is $\lambda = \lambda_{max}/2$ (see text).

applied to walls and ceilings; however, it must be tested for its absorption, even if it is made from hard material (Fujiwara & Miyajami 1992).

The quadratic residue diffuser, shown in Figure 9.13, diffuses over a frequency range that can be characterised by the wavelengths λ_{max} (lowest frequency) and λ_{min} (highest frequency). The well width is

$$W < \frac{\lambda_{min}}{2} \qquad (9.3)$$

and the well depth is

$$d_n = \frac{s_n}{N} \frac{\lambda_{max}}{2} \qquad (9.4)$$

where $s_n = \mathrm{Res}(n^2 \bmod N)$ (quadratic residues s_n are numbers taken as the least non-negative residues with modulus N), $n = 0, 1, 2, \ldots,$ and N is a prime number. For example, when $N = 17$, starting with $n = 0$, the sequence is

$$s_n = 0, 1, 4, 9, 16, 8, 2, 15, 13, 13, 15, 2, 8, 16, 9, 4, 1$$

so that N is the period.

Various materials for diffusers can be considered as follows: for reflective purposes, cement and tiles are acceptable, while plywood and other boards are used for absorption in the low-frequency range. Further enhanced absorption is obtained with porous materials and perforated boards, which provide particular absorption characteristics. These devices are often used to modify the reverberation characteristics. Figure 9.14 shows examples of measured data.

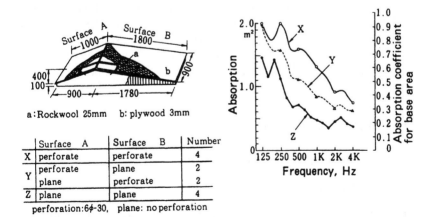

	Surface A	Surface B	Number
X	perforate	perforate	4
Y	perforate	plane	2
	plane	perforate	2
Z	plane	plane	4

perforation: 6ϕ-30, plane: no perforation

Figure 9.14 Absorption measured on four pieces of a prismatic diffusing element (Maekawa *et al.* 1964).

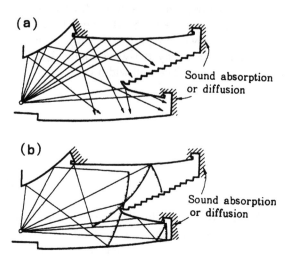

Figure 9.15 Geometrical drawing investigations of room section profiles: (a) sound-ray tracing; and (b) wave-front drawing.

C. Geometrical drawing study of room shape

Once a room has been designed with due reference to the components described above, the whole room shape can be developed as a functional acoustic system. There is a geometrical method of representing the behaviour of sound transmission and reflection, as illustrated in Figures 9.1– 9.10 and also Figure 1.6(b) in Chapter 1. First, in order to distribute the prime reflections uniformly over the whole audience area, avoiding undesirable phenomena such as echoes, it is necessary to adjust the positions and angles of the reflecting surfaces. Further, in order to avoid the creation of long-path echoes and sound foci, etc., a trial-and-error method is employed to adjust the diffusive and absorptive treatment until an adequate reflected sound distribution is obtained. Figure 9.15 shows an example of the investigation of the longitudinal section of the room: (a) with sound rays; and (b) with wave fronts. The same analysis should, of course, be applied to the floor plan.

9.3 Planning the reverberation

A. Determination of optimum reverberation time

a. Room volume and purpose

With a knowledge of the major purpose for which a room is to be used and its volume, the optimum reverberation time and its frequency characteristics can be found from Figures 3.22 to 3.24 in Chapter 3.

In the case of a multi-purpose hall, the programme of events should be obtained and the architectural design should then be dictated by the bias towards, for example, music, or, alternatively, drama, etc.

b. Variable reverberation time facilities

In a multi-purpose room, the requirement is for the reverberation to be variable. Figure 4.31 shows an example of devices for achieving this flexibility. However, the area which can be used for such a variable surface function is often limited, whether it is on the ceiling or walls, resulting in variations up to 10–20%. Further, since the timbre or sonority is also changed by changing the frequency characteristics of absorption or is distorted for unknown reasons, it is difficult to cover the range shown in Figure 3.22 with complete ease. A reasonable approach to the problem is to install a thick curtain with a deep air space behind it.

Currently, electro-acoustic devices are being developed that can actively control the sound field and increase the reverberation time over a wider range. Even if the reverberation time could be made totally variable, either architecturally or electro-technically, it is doubtful whether anyone could be found who would be capable of operating such a device so that it would always produce the best acoustic conditions. There is even the possibility of making them worse. In order to make best use of the facilities, a new monitoring system capable of indicating the necessary operating conditions based on objective measurements would be required.

c. Electro-acoustic systems

It is of fundamental importance to note whether an electro-acoustic system acts as the main facility or takes on a subsidiary, supporting role. In recent years, although microphones and loudspeakers are frequently used in many places, there have been many unsuccessful applications due to improper design of the system or difficulty of operation. If an electro-acoustic system is to be given priority, then it is advisable to adopt a somewhat shorter reverberation time than the recommended value indicated in Figure 3.22, so as to reduce operational difficulties that can arise, because recorded sounds already contain the proper reverberation. This approach also suppresses howling. Reverberation to reinforce stage sound can simply be added with an electronic reverberator (Chapter 10). At the very least the sound absorption surrounding the microphone should be increased.

B. Determination of room volume

The reverberation time is directly proportional to the room volume and inversely proportional to the total absorption as shown in Equation (3.26). Since an audience has a large absorption, unless the unit volume/person

Table 9.1 Recommended room volume per seat

Concert hall	8–10 m³
Opera, multi-purpose hall	6–8 m³
Theatre, cinema	4–6 m³

for the space is appropriate, the reverberation time may be inadequate. The unit-floor area per seat is about 0.6 m² including aisles, and, by adjusting the ceiling height, the unit volume should meet the value indicated in Table 9.1, which shows the recommended unit volume for a variety of rooms. Any room in which music is performed needs a larger volume because of the need for a longer reverberation time; however, too large a volume decreases the acoustic energy density, resulting in unfavourable acoustics. On the other hand, a movie theatre requires a smaller volume, hence, a shorter reverberation time because the use of an electro-acoustic system is necessary. Of course the smaller the volume, the more economical the construction and the lower the running costs.

C. Sound absorption design

a. Determination of necessary absorption

Once the target values of reverberation time T and room volume V are determined, the necessary absorption A can be obtained from the reverberation time formula, Equation (3.38) or (3.37), from which the absorption of the audience is subtracted. This result is then the basis for the design, construction and finishes of walls and ceilings.

b. Effect due to variable audience

For an auditorium whose absorption is largely affected by the audience, the reverberation times, both when fully occupied and empty, should be calculated; the seats should provide absorption that is close to that of the full audience, in order to minimise the effect due to the different occupancy. However, it should be noted that the seats must not have too much absorption when a member of the audience sits on it.

c. Location of absorbents

Having designed the room shape, the following steps are then followed: (1) Determine the absorption required for local spot surfaces that need treatment in order to avoid echoes and flutter. (2) Apply reflective (live) treatment around the stage and absorptive (dead) treatment along the rear wall. These are called the live-end and dead-end methods. (3) Furnish dead treatment around microphones when they are used as electro-acoustic devices, and (4) distribute the absorbents in patches of small dimensions, comparable to sound wavelengths, instead of using an absorption treatment over a large extended area, in order to diffuse the sound field in the room and to increase the total absorption by the area effect (Figure 4.10).

d. Remarks on calculation

Following on from the above guidelines, the frequency range from 125 to 4000 Hz should be investigated and the final reverberation times calculated using Equation (3.38), satisfying oneself that, by a process of trial and error, the optimum values have been attained by selection from various combinations of materials and construction details (Chapter 4 and Table A.1). Table 9.2 shows an example of this type of calculation.

9.4 Computer simulation and acoustic model analysis

A. Computer-aided analysis of the sound field in a room

a. Computer-aided design based on geometrical acoustics

For the simulation of sound in large rooms, there are two classical geometrical methods, namely the ray tracing method and the image source method. For both methods, there is a problem in that the wavelength or the frequency of the sound is not inherent in the model. This means that the geometrical models tend to create high-order reflections, which are much sharper than would be possible with a real sound wave. One way of introducing the wave nature of sound into geometrical models is by assigning a scattering coefficient to each surface. In this way, the reflection from a surface can be modified from a pure specular behaviour into a more or less diffuse behaviour, which has proven to be essential for creating reliable results from computer models.

b. The ray tracing method

The ray tracing method uses a large number of particles, which are emitted in various directions from a source point. The particles are traced around the room losing energy at each reflection according to the absorption coefficient of the surface, see Figure 9.16. When a particle hits a surface it is reflected, which means that a new direction of propagation is determined, e.g. as a specular reflection according to Snell's law (see Equation (1.26)) or in a more or less random direction taking scattering due to surface roughness into account.

Table 9.2 Reverberation time calculation for a concert hall

Places	Finish materials	Area S(m²)	125 Hz α	125 Hz Sα	250 Hz α	250 Hz Sα	500 Hz α	500 Hz Sα	1 KHz α	1 KHz Sα	2 KHz α	2 KHz Sα	4 KHz α	4 KHz Sα
Floors	Vinyl chloride tile on concrete	494	0.01	5	0.02	10	0.02	10	0.02	10	0.03	15	0.04	20
	45-mm timber flooring on joists	118	0.15	18	0.10	12	0.08	10	0.07	8	0.05	6	0.05	6
Ceilings	12-mm gypsum board with large air space	431	0.25	108	0.15	65	0.10	43	0.08	34	0.06	26	0.05	22
	Diffuse and absorptive construction	92	0.30	28	0.30	28	0.30	28	0.30	28	0.30	28	0.20	18
Walls	Mortar, paint	403	0.01	4	0.02	8	0.02	8	0.03	12	0.03	12	0.03	12
	24-mm gypsum board with 150 mm air space	60	0.18	11	0.13	8	0.06	4	0.06	4	0.06	4	0.05	3
	Perforated board, 25-mm glasswool, large air space	92	0.55	51	0.80	74	0.75	69	0.48	44	0.33	30	0.15	14
	Reflecting panels on stage	321	0.15	48	0.15	48	0.10	32	0.08	26	0.07	22	0.06	19
Door	Vinyl leather with upholstery	44	0.10	4	0.15	7	0.20	9	0.25	11	0.30	13	0.30	13
Window	Glass panes	16	0.30	5	0.20	3	0.15	2	0.10	2	0.06	1	0.03	0.5
Chairs	Theatre chair upholstered	660	0.15	99	0.20	132	0.28	185	0.30	198	0.30	198	0.30	198
Persons	Adult audiences	660	0.20	132	0.25	165	0.33	218	0.40	264	0.40	264	0.40	264
Surface	$S = 2071\,\text{m}^2$													
Volume	$V = 4850\,\text{m}^3$													
Total absorption (m²)	Unoccupied condition			380		393		399		376		355		325
	With full audiences			413		426		432		442		421		391
Air absorption (4mV)										17		43		109
Reverberation time (s)	Unoccupied			1.86		1.79		1.76		1.80		1.81		1.69
	With full audiences			1.70		1.64		1.61		1.52		1.52		1.44

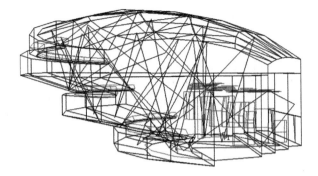

Figure 9.16 Ray tracing from a source position in a computer model of a room. Only one single ray is shown.

c. The image source method

The image source method is based on the principle that a specular reflection can be constructed geometrically by mirroring the source in the plane of the reflecting surface, see Figure 5.7.

The advantage of the image source method is that it is very accurate, but if the room is not a simple rectangular box there is a serious calculation problem, because the number of possible higher-order image sources increases exponentially with reflection order. For this reason, image source models are only used for simple rectangular rooms or in such cases where low-order reflections are sufficient, e.g. for the design of loudspeaker systems in non-reverberant enclosures. An example of reflections up to the third order between a source and a receiver in a concert hall is shown in Figure 9.17.

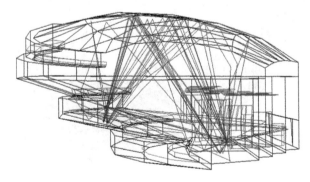

Figure 9.17 Calculated reflections up to the third order between a source point and a receiver point in a room.

d. Hybrid methods

The disadvantages of the two classical methods have lead to the development of hybrid models, which combine the best features of both methods.

One of the advantages of a room acoustic computer model is the possibility of calculating the response at a large number of receivers distributed in a grid that covers the audience area. An example is shown in Figure 9.18. It can be useful for the acoustic designer to see a mapping of the spatial distribution of acoustic parameters. Uneven sound distribution and acoustic weak spots can be localised and appropriate countermeasures can be taken.

e. Auralisation

It is an old idea that it might be possible to listen to sound in a room by a simulation technique using the impulse response from a room model (Spandöck 1934); however, it is only in recent years, with the development of computer technology, that this idea has developed to be a useful tool for room acoustic design. The technique, to make a room model audible, is called auralisation (in analogy to visualisation). The auralisation technique offers the possibility of using the ears and listening to the acoustics of a room during the design process. Several acoustic problems in a room can easily be detected with the ears, whereas they may be difficult to express with a parameter that can be calculated.

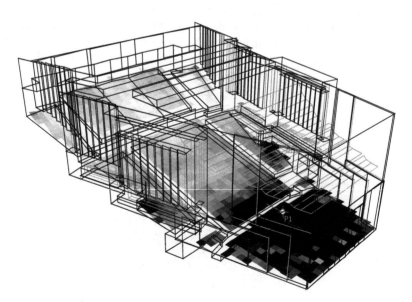

Figure 9.18 Mapping of acoustic parameters calculated in a grid that covers the audience area in a concert hall.

In principle, it is possible to use impulse responses measured in a scale model for auralisation. However, the quality may suffer seriously due to non-ideal transducers. The transducers are one reason that the computer model is superior for auralisation. Another reason is that the information about each reflection's direction of arrival allows a more accurate modelling of the influence from the listener's head.

The auralisation options in a typical room acoustic computer programme are based on binaural technology, allowing a three-dimensional presentation of the predicted acoustics through headphones. At the receiver point, the impulse response is calculated for each ear of a listening person. An example is shown in Figure 9.19. The listening signal is an anechoic recording, which can be speech, song, music, or whatever could be relevant for a listening test. When the anechoic signal is convolved with the impulse response the result is a sound signal, which, when listened to, gives the illusion of listening to the sound in the room that has been modelled.

f. Sound field analysis using wave theory

The analysis of room acoustics by wave theory is described in Chapter 3, but is limited to simple cases. However, due to the growth in computer

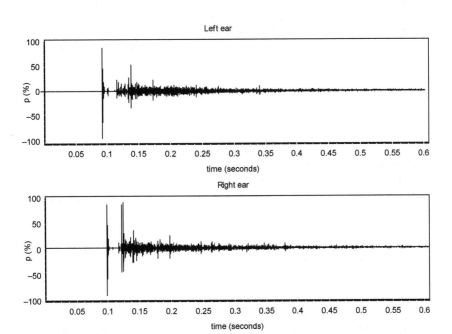

Figure 9.19 A calculated pair of impulse responses (for left and right ear) that can be used for auralisation. In this example, the calculated impulse response was 2 s long, but only the first 600 ms are shown.

technology and through the development of solutions to the wave equation based on integral equation analysis of large spaces, using the finite element method or the boundary integral equation (Sakurai 1987; Terai & Kawai 1990), we are now in a position to apply such mathematical techniques to the practical design of room acoustics.

Wave models are characterised by creating very accurate results at single frequencies, in fact often too accurate to be useful in relation to auditoria and similar environments, where results in octave bands are usually preferred. Another problem is that the number of natural modes in a room increases approximately with the third power of the frequency (see Equation (3.17)), which means that the usual wave theoretical models are restricted to low frequencies and small rooms due to the calculation time involved.

However, with the new FDTD (finite-difference time-domain) method for the calculation of impulse responses, it has become possible to solve the sound field in a large hall and to calculate the room acoustic parameters, such as shown in Table 3.3 in the mid-frequency band 500–1000 Hz (Sakamoto *et al.* 2008).

B. Acoustic model analysis

Changing boundary conditions is so easy in computer simulation that it is a useful method of investigating room shapes at an early phase of design. At the detailed design stage, however, acoustic measurements using a scale model are recommended to determine the shape and material needed for the interior design. Historically, many methods have been used in model studies that relate to the scale ratio.

a. Ripple tank method (scale about 1/50)

In place of sound, ripples in a shallow water basin are used; the wavelength range to be handled is rather narrow to serve as an analogue of the sound wave and only the wave motion in two dimensions can be seen. Three-dimensional space can not be studied by this method; therefore, it is not used.

b. Light ray method (scale 1/50–1/200)

Instead of sound, light rays are an effective means of reducing the problems encountered in a geometrical drawing with a complicated shape. This method is often used to examine reflections in two dimensions and to investigate the sound intensity distribution (using luminous intensity) in three-dimensional space (see Lit. Al). Although confined to geometrical acoustics, new detailed investigations have been reported (Pinnington & Nathanail 1993).

c. Ultrasonic wave method

Using a 1/30–1/8 scale model in which a spark pulse or a small speaker system generates a sound, whose wavelength is of the same scale, the impulse response, sound pressure distribution, echoes and reverberant decay can be measured using small microphones. Applying this technique it is possible to investigate not only room shapes but materials for the room boundaries. To make a wavelength $1/n$ of the original means that the frequency must be n times larger than the original, i.e. in the supersonic range. Measuring instruments in that frequency range have become sufficiently well developed.

The materials for the model are obtained by looking for other materials with an absorption coefficient appropriate to the scale frequency, using measurements in a scale reverberation chamber; they are, of course, different from those building materials used in the actual room. Resonant absorbers may be designed in the normal way; porous absorbers tend to have a higher density and window panes can be simulated with thin panels made from aluminium and so on.

One major problem is that the attenuation factor m due to air absorption (see Equation (3.38)) is very large in the ultrasonic range. Although the measurements are generally used for the investigation of room shape, even ignoring the above effect, the air absorption coefficient m must be n times larger in the $1/n$ scale model. Air absorption is due both to molecules of oxygen O_2 and moisture H_2O. However, since no absorption occurs in the absence of oxygen and moisture, it seems reasonable to apply a method using either nitrogen or dry air. So far, a method using dry air, whose relative humidity is 2–3%, has been used, although the effect on the model materials and instruments is still of concern (Brebeck *et al.* 1967). But, by using nitrogen, quite a good approximation to the air absorption is obtained in a 1/10 scale model, though there is a possible risk of oxygen shortage for operators (Ishii & Tachibana 1974). Therefore, if these problems are solved, the sound field can be realised in a 1/10 model both in the frequency and time domains.

Then, by using small microphones in both ears of a 1/10-scale dummy head, the impulse responses are measured in the model and multiplied by 10 in the time scale in order to obtain the equivalent responses in the real room, so that more precise binaural auditory tests can be performed by convoluting dry music, recorded in an anechoic chamber, with both impulse responses, using a computer (Els & Blauert 1986). Figure 9.20 shows an example of a real concert hall and the 1/10 scale model, which was applied for acoustic measurements in order to confirm the design. This is an example of a very detailed scale model, which was not only for acoustic evaluations, but also for visual purposes.

When the model scale is smaller than 1/20, useful information can still be obtained by objective testing, such as observing the echo time pattern,

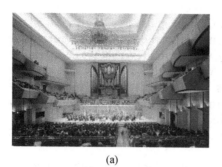 

(a)   (b)

Figure 9.20 The Yokohama Minato-Mirai Concert Hall (a), and the experimental 1/10 scale model of the same (b). Reproduced with permission from H. Tachibana (2010).

for obtaining a uniform sound distribution and smooth decay curves at any position in the room (Barron & Chiney 1979). It may also be possible to theoretically correct the effect of air absorption using a computer.

9.5 Practical examples of the acoustic design of rooms

Some examples that present acoustic interesting problems have been selected from the field of architecture and building.

A. School classroom

When the number of students is less than 100, a rectangular room shape is acceptable, bearing in mind the three-dimensional proportions (see Section 9.2.A.a). If the number is greater than 100 or 150 say, the room section in Figure 9.21 is preferred. As this figure shows, if the first reflected sound is used properly, no sound reinforcing system is necessary since the audience is provided with sufficient speech intelligibility, so long as the background noise is low and there is a slightly longer reverberation time. However, when an audio-visual facility is required, or a sound reinforcement system is used with a rather high background noise level, then it would be preferable to apply some absorptive treatment to the rear wall.

B. Gymnasium

Since a gymnasium usually has a large volume with relatively small seating area, the reverberation time tends to be very long and yet the space is frequently used as a multi-purpose hall, thus, it is usually difficult to provide satisfactory acoustics. Moreover, the use of an inappropriately designed electro-sound system adds to the problem as speech intelligibility

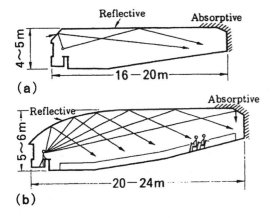

Figure 9.21 Design of longitudinal section of class rooms: (a) for 150–300 seats; and (b) for 300–500 seats.

may become considerably worsened due to the overlapping effect of delayed sound from distributed loudspeakers.

Since the installation of a sound reinforcement system is indispensable in a room of large volume, the system design should be carried out according to the recommendation in Chapter 10. At the architectural design stage, sufficient absorption should be provided to make the reverberation time as short as possible. For example, resonance absorption based on a slotted wainscoting and absorbent ceiling with glass fibre or rock wool are effective. Since the floor surface is always reflective, the ceiling and other surfaces cannot have too much absorption. At the same time, the room plan and section should be closely examined to avoid any possible defects that could arise from parallel walls or domed ceiling and also from flutter echoes as shown in Figure 9.3.

C. Theatre

There is a wide range of theatre designs possible, depending not only on the type of play or drama, but also on the style of production and direction. An overriding influence tends to be the priority given to visual rather than acoustic requirements. Therefore, the seating area is limited in distance from the stage and is often extended in a fan shape (Figure 9.10), so that the shape of walls and ceiling must be treated as an essential acoustic component. Although actors require the natural voice to be intelligible, an electro-acoustic system will be used for background sound, and because there may be musical performances, it is also important that the reverberation time is chosen from the middle of C and D in Figure 3.22 in Chapter 3. As regards the electro-acoustic system, the loudspeaker installation should be such that the audience is not aware of its existence.

D. Concert hall

This is a hall whose main objective is the performance of music. Some plans and sections of historically well-known halls are shown chronologically in Figures 9.22 and 9.23, respectively.

Part (a) in Figures 9.22, 9.23 and 9.24 shows a representative hall built in the nineteenth century, which, because of its rectangular shape, is referred to as a shoebox, with a rich sonority for many people. This basic design has been adopted all over the world; the one shown in Figures 9.20 and 9.22(e) is intended to be a descendant of it. In contrast to its end stage, an arena stage is possible, as shown in Figure 9.22(c). This is a descendant of the classic theatre in Greece; Figure 9.22(d) shows a more faithful version of it. As for the directivity of musical instruments, the acoustic quality in the seating area must be changed according to its orientation with respect to the

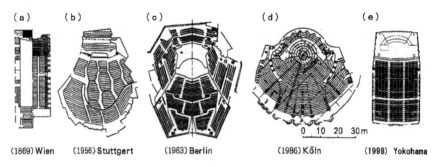

(a)	(b)	(c)	(d)	(e)
(1869) Wien	(1956) Stuttgart	(1963) Berlin	(1986) Köln	(1998) Yokohama

Figure 9.22 Examples of concert-hall plans.

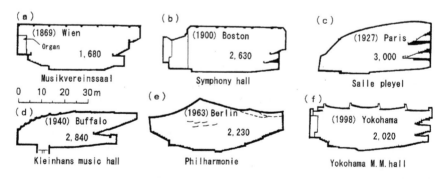

Figure 9.23 History of longitudinal sections of concert halls, showing seating capacity.

Figure 9.24 Grosser Musikvereinssaal Vienna.

stage. Recently, halls of this type have been built throughout the world and may have been influenced by a desire for the performance to have a visual as well as a musical impact.

Note the longitudinal sections in Figure 9.23. At the start of 'modern architectural acoustics' in 1900, a parabolic ceiling as in Figure 9.23(c) was considered desirable as a means of effectively collecting sound from the stage and sending it to the audience but it creates problems by gathering the audience noise and concentrating it back on the stage.

In the next example Figure 9.23(d), it was decided that acoustic quality could be improved by introducing many more reflections from surfaces divided into small diffusing elements, rather than reflections from a larger continuous shallower concave surface, thus improving natural brilliance. The room shape of Figure 9.23(e) was investigated by a 1/8-scale acoustic model, where reflecting panels, called clouds, were suspended above the stage where the ceiling became too high to reflect the stage sound with an appropriate time delay. In the photograph (Figure 9.25), we can see the ceiling reflectors and also the seating area, which is divided into blocks like a terraced field. This is referred to as 'vineyard' seating and is intended to supply lateral reflections from the terrace walls, which would otherwise be in short supply, caused by the rather wide plan as seen in Figure 9.22(c) (Cremer *et al.* Lit. B27).

As regards Figure 9.23(f), it would appear that the ceiling section is the successor to that in Figure 9.23(d) and that the seating arrangement follows that in Figure 9.23(c).

Now, as an important page of history, we must not forget the New York Philharmonic Hall in the Lincoln Centre, shown in Figure 9.26, which was completed in 1962, a year which also saw the publication of the well-known book by Beranek (Lit. A4). After several modifications, all of which did nothing to enhance its reputation, the Hall disappeared and was replaced

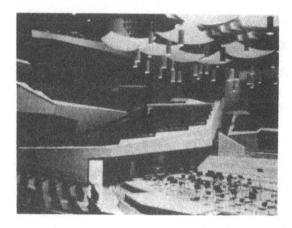

Figure 9.25 Neue Philharmonie Berlin.

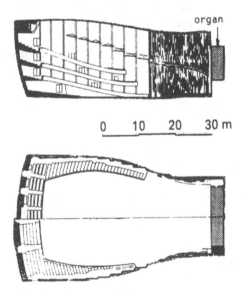

Figure 9.26 New York Philharmonic Hall (1962–1975).

by the Avery Fisher Hall of the shoebox-type in 1976. During its short life, researchers were excited at discovering the reason for the failure and, as a result of experiments, great strides were made in room acoustics (Lit. A11). Recently, computer simulation and acoustic-model study have shown marked progress. However, while the exact prediction of room acoustics is still not easy, the creation of acoustic quality is gradually becoming achievable through a scientific approach to the art of musical performance.

E. Multi-purpose hall

Public halls, like civic auditoria and most commercial rental halls, are for multi-purpose use, so that the reverberation time has to be adequate for both music and speech, but priority must be given to the requirements of the most frequent performance, although, unfortunately, it is not easy to anticipate the true requirements. Once there was an inclination towards shorter reverberation times for the purpose of broadcasting, but nowadays almost everybody in Japan aims at concert hall acoustics. This seems to be rather a quirk of fashion, as concerts, especially in the provinces, are rare.

The style of an auditorium should meet the requirements of the social activity in the locality in which it is situated. In Japan, most multi-purpose halls have a stage for drama and fixed seats for the audience, and a concert stage with a movable reflecting panel for an orchestra shell (Figures 10.2, 10.3 in Chapter 10). In Europe, there is usually a fixed concert stage, an organ and movable seats on a flat floor for the purpose of banquets and balls, etc. Figure 9.27 shows a European example, which has 1400 seats for concerts.

As for having a variable reverberation time, there are various problems as described in Section 9.3.A, but, without any electronic hardware, the difference between a stage with an orchestra shell and a drama stage where the reflecting panels have been removed is about 20–30% when using a highly absorptive treatment of the walls and ceiling in the stage area. When the auditorium is not so large (up to 500–600 seats), a curtain 2–3 m high above the floor improves speech intelligibility, even if the variation in reverberation time is small.

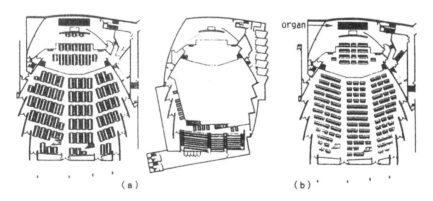

Figure 9.27 Variable seating in Beethovenhalle in Bonn (seating capacity 1407): (a) banquet style with 960 seats; and (b) parliamentary style with 492 seats.

Additionally, active control by electro-acoustic devices would be useful for changing the acoustic conditions of a hall to suit any performance.

PROBLEMS 9

1   When designing the shape of a room, what is the fundamental strategy from the acoustic point of view?
2   List the important acoustic points, starting from the basic design of a concert hall, giving the sequence.
3   When designing a multi-purpose hall, how do the acoustic requirements differ from a concert hall?
4   What acoustic problems should be given special consideration in the architectural design of the following rooms? (1) Conference room (about 40 people); (2) small auditorium (approximately 500 seats); (3) banquet room (about 200 people); (4) gymnasium in a primary school; (5) drawing office in a factory; (6) large office (about 30 people).

10 Electro-acoustic systems

Now we have reached the stage where there are no building types that do not possess electro-acoustic systems. Also, in the open air, electro-acoustic systems play an important roles as information services. An architectural or environmental planner must, therefore, have a basic knowledge of these systems sufficient to enable them to cooperate with audio engineers.

In this chapter we shall discuss several items that are closely related to the basic scheme for both the architecture and the environment, especially the sound reinforcement system, which is the most important equipment as far as planning the electro-sound is concerned.

Another purpose of this chapter is to give basic knowledge to students for acoustic experimental work.

10.1 Functions and aims of electro-acoustic systems

A. Sound reinforcement system

This system consists of a microphone, amplifier and a loudspeaker, usually situated in the same acoustic environment. The microphone receives speech or music signals and the loudspeaker radiates the amplified sound. The system is useful for obtaining high articulation and intelligibility by raising the loudness when the sound pressure level of the original signal is insufficient, or the SN ratio is too low due to background noise. The term **SR** (Sound Reinforcement) implies powerful radiation at high level, especially for modern music, and the term **PA** (Public Address) implies amplified speech, though there is no strict dividing line.

A major problem is that of **howling**. This is an acoustic phenomenon, which totally swamps the ability to hear due to an oscillatory sound caused by feedback, which happens when the sound radiated from a loudspeaker is picked up by the microphone. How to control it and still amplify up to the required sound level without oscillation, is the most important problem, not only for electro-acoustic design, but also for room acoustic design.

B. Sound reproducing system

This system has additional means of reproducing recorded media (tape or disc and others), but the original sources are in a different space from the listeners as with radio, or cable transmission or sound effects in a drama theatre or cinema. There is no howling, so that it is much easier to obtain good quality with appropriate volume. In a public space, however, the sound system is seldom of a high enough quality for the transmission of information, and, sometimes, regrettably, even gives rise to environmental noise pollution, owing to poor equipment and/or inexpert operators.

C. Sound recording system

This system records the sound signals of musical performances, and also of lectures and conference presentations. It is most important that the recording apparatus is carefully selected since there is a wide range of sophistication, from a level of simple toy to a professional recorder for manufacturing commercial recorded media.

D. Auxiliary facility

There are several pieces of equipment attached to the sound reinforcement and reproducing systems with facilities for actively controlling the sound field. The time delay devices and reverberation equipment owe much to the rapid development of digital technology. Table 10.1 shows as a reference, important electro-acoustic equipment, categorised according to application, but in the following sections it will be useful to look at it from the point of view of architectural planning and design.

E. Public security system for emergencies

Almost all buildings require a public address and alarm system for emergency use, for alerting people to accidental events such as fires or earthquakes, etc. The function of these systems was originally as sound reproducing systems; however, their installation is regulated, with quite delicate technical standards. Therefore, when a security system is needed, all work should not only follow this book but also the regulations of the local government.

10.2   Outline of sound reinforcement systems

A. Functions of loudspeakers

a. Proper layout of loudspeakers

Generally the most essential problem is planning the layout of the loudspeakers in the room or given space. The aims are as follows:

Table 10.1 Principal equipment for electro-acoustic systems

Instruments	Purpose	Reinforcement system			Reproducing system		Recording system	
		Lecture	Conference	Music	Announcement	Effect Sound	Music	Speech
Input system	Microphones	○	○	○	○		○	○
	Wireless microphones	○	○	○				○
	Rise Mic. with hydraulic jack	○	△	○			△	
	3 point (hanger) of mic.		△	○			○	
Control system	Record players				○	○		
	Record & playback device	○	○	○	○	○	○	○
	Mixing console	○	○	○	○	○	○	○
	Time-delay machines	○	△	○	△	○	△	
	Reverb. Machines			○		○	○	
	Sound image control	△	△	△	△	○	△	△
	Effects console					○	△	○
	Record & playback device	○	○	○	○	○	○	○
Output system (loudspeakers & amplifiers)	Audience area — Centralised system	○	△	○	○			
	Distribution system	△	○	△	○	○		
	On the stage — for Music			○				
	Stage monitor	○	△	○				
	Lobby, Foyer	○	○	○	○	○		
	Monitor in control room	○	○	○	○	○	○	○
Simultaneous interpretation system			○				○	○

○ : necessary,   △: need to consider

Reproduced by permission of Kyouritsu-Shuppan Co. Ltd.

1   Uniform distribution of sufficient loudness for all listeners.
2   Natural directional perception of the original source, without giving the impression that the sound is coming from the loudspeakers.

In addition, it is most important to control the accompanying echo and howling, because when they happen, not only is all effort wasted, but they are very unpleasant for the audience.

b. Basic functions required of loudspeakers

The first work in planning an electro-acoustic system is the selection of loudspeakers available in the market. There are many manufacturers and many types of loudspeaker. The ideal requirements would be as follows:

1   Good tone quality, as natural as the original sound source.
2   Exact directivity that just covers the defined service area.
3   Supplying sufficient sound power for the whole audience, without affecting other areas.

These requirements are ideal and are also technological goals. At present, we cannot completely obtain them. But, we have to approach the goals.

B. Radiation patterns of sound sources

In the beginning of Chapter 5, we learned how to calculate the attenuation of radiated sound due to distance for three types of sound sources. In electro-acoustic systems, a single loudspeaker is treated as a point source. Loudspeakers set in an array in line, work as a line source. And loudspeakers placed on a wide plane, work as a plane source as shown in Figure 10.1.

We must consider the directivity of the sources and the attenuation due to the distance from the radiating source.

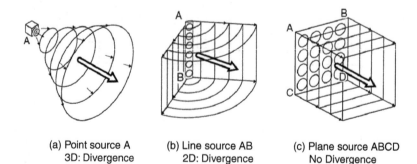

| (a) Point source A | (b) Line source AB | (c) Plane source ABCD |
| 3D: Divergence | 2D: Divergence | No Divergence |

Figure 10.1 Schematic sound radiation patterns for three types of loudspeaker arrangements.

C. *Layout of loudspeakers in a room*

a. *Centralised system*

One or two loudspeakers, having proper directivity, should be situated as close to the original source as possible. This arrangement is commonly used in auditoria, lecture halls and gymnasiums, etc., of modest scale. Its merits are as follows:

1   good localisation of the source is easy to obtain for the listeners;
2   prevention of echoes is fairly simply achieved by controlling the directivity, so that not too much direct sound is incident on the ceiling or walls, but only on the audience, and also by sound absorptive or diffusive treatment on the rear wall;
3   speech intelligibility can be raised by the amplified direct sound, while making the virtual reverberation time short by means of the proper directivity of the loudspeaker arrangement described above;
4   the likelihood of howling can be more or less totally overcome by considering both the directivities of the microphone and loudspeakers, and selecting their positions accordingly; in practice attention should be given especially to the microphone position.

b. *Distribution system*

When the same signal is fed to two loudspeakers some distance from each other, a listener sitting equidistant from the two loudspeakers will perceive a sound image that appears to be at the midpoint between them. But, if the distances are different, and also when there are many loudspeakers, the sound image appears to be situated at the nearest loudspeaker, by the **law of the first wave front** (see Chapter 1, Section 1.12), and the other loudspeakers seem to be silent. When the difference in arrival time is in the range of several to slightly more than 20 ms, colouration is caused by

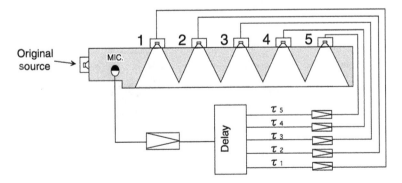

Figure 10.2 Basic principle of distributed loudspeakers with a time-delay system for controlling the localisation of a sound source.

phase interference, and when it is more than 30–50 ms, echoes will occur (Chapter 3, Figure 3.29).

As shown in Figure 10.2, when the same signal is fed to many distributed loudspeakers, the sound pressure distribution of continuous noise is uniform in the steady state. However, if the signal is meaningful sound, such as speech or music, the many different time delays of the arriving signals will interfere with each other in a complicated way and reduce the clarity of the sound. Sometimes listeners cannot understand anything at all. Therefore, the indispensable fundamental design parameters in a distributed loudspeaker system are as follows:

1   The service area of each loudspeaker must be limited in terms of its directivity, and the sound level should be controlled in order not to disturb adjacent areas. Clarity will be enhanced by this procedure.
2   The signal to each loudspeaker should be given a time delay, using an electronic delay, as shown in Figure 10.2, so each listener receives sound from the nearest loudspeaker a few milliseconds after the arrival of the direct sound from the original source.

This localises the sound image in the original source direction by the law of the first wave front and gives the listener good directional perception of high quality. Figure 10.3 shows an example of the type of ceiling loudspeaker used in Figure 10.2.

c. Combination of centralised and distribution layout systems

In practice, the design layout of the loudspeaker system depends on the size and shape of the room, as well as the specific purpose for which it is intended. Generally, it is recommended that a centralised system is used, with distributed loudspeakers added at spots where there is a poor supply of sound. In this case, also, deliberate consideration needs to be given to the control of the defects in the distribution system described above.

(a) Woofer 25cm, + Horn coaxial,

Angle: 60° × 60°,

Program input: 180 w.

(b) Front grill at ceiling.

Figure 10.3 Example of a ceiling speaker.

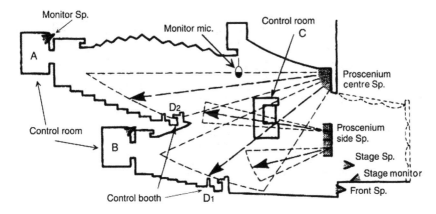

Figure 10.4 Loudspeaker layout and sound control rooms in a multi-purpose hall.

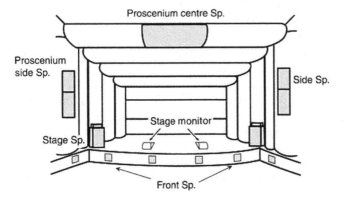

Figure 10.5 Loudspeaker layout around the stage in a multi-purpose hall.

The example shown in Figures 10.4 and 10.5 has a main loudspeaker **Centre Sp** at the centre of the proscenium, intended to amplify the sound of the original sources on the stage for the benefit of the whole audience.

The loudspeaker system is generally installed in a large cabinet weighing several tens of kilograms. When it is hung from the ceiling, proscenium, etc., it is important that the loudspeaker is separated from its surroundings and supported by means of a vibration isolating rubber suspension system.

Figure 10.6 shows an example of hidden loudspeakers in the proscenium, which has two kinds of loudspeaker units, called **two way**, one for the low-frequency range called a **woofer** and the other for mid and high frequencies called a **tweeter**, and two systems, one for near and the other for distant audiences.

The loudspeaker should be fully exposed to the auditorium so as not to impair the sound quality. If, for aesthetic reasons, it is necessary to hide it in

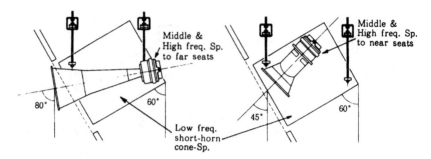

Figure 10.6 Suspending the proscenium loudspeaker.

the ceiling or walls, the front face should be as open as possible and covered with a sound transparent coarse cloth and should not be obstructed with either a perforated board or dense grill. Moreover, for ease of maintenance, the loudspeaker should be constructed so that it can be lowered to the floor or accessible from a catwalk, etc.

Figures 10.7 and 10.8 show an example of the stage speaker and stage monitor shown in Figures 10.4 and 10.5.

The proscenium side speaker, shown in Figures 10.4 and 10.5, is almost the same style as the speaker in Figure 10.6. In recent trends, however, both low- and high-frequency range units are set in a smaller box, whose shape is convenient to assemble into an array or cluster, and hanging or stacking on the stage, as shown in Figures 10.9–10.11.

The additional loudspeakers in Figure 10.5 are distributed at dark spots, i.e. **Front Sp** at the stage front for the front seats, and a speaker on

Top: Constant directivity horn, H:90°×V:40° above 800Hz
+Woofer 30cm: 65~800Hz, Program Input: 150W+600W

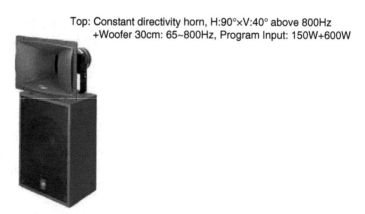

Figure 10.7 Example of a stage speaker.

Woofer 30 cm + Horn, H:90° × V:40°,
Program input: 360 W.

Figure 10.8 Example of a stage monitor.

the stage for the performers, indicated as **Stage Monitor** (Figure 10.8.). The proscenium side speakers, indicated by **Side Sp,** are used not only to reproduce stereo music, but also to control the sound image of the centre speaker, to shift it down to a lower position in combination with the front speakers.

For sound effects in drama and cinema, many loudspeakers may be distributed in the walls and ceiling or floor. In this case, it is necessary to have a skilled operator to control the **effects console**. Therefore, deliberate planning is required, taking into account the purpose, frequency of performance and technical ability of the operator employed by the hall.

Figure 10.9 Example of a line array speaker, suspended in a concert hall. Top box shows a sub-woofer, lower than 80 Hz.

Figure 10.10 Line array speaker. Woofer 30 cm, cone and horn tweeter. Crossover: 1000 Hz. Directivity: H: 90° × V: 5–15°/unit. Program input: 450 W/unit.

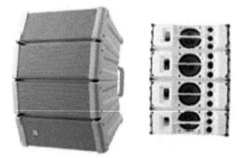

Figure 10.11 Compact array speaker. Woofer 12 cm, cone × 4 and dome tweeter × 12. Crossover: 4 kHz. Directivity: H: 100° × V: variable = 15°, 30°, 45°, 60°, depending on assembly style. Program input: 450 W/unit. Right side: without front net.

Figure 10.12 Line-array two-way speaker, 70–20 000 Hz. Crossover: 3500 Hz. Program Input: 600 W/unit.

d. Application of line source and plane source

Where speech is the main purpose, like a lecture hall, the line-array (column) loudspeaker that consists of many loudspeaker units, as shown in Figure 10.12, is frequently used. The directivity does not spread vertically as shown in Figure 10.1(b), and in the horizontal plane has the same pattern as a single loudspeaker. Therefore, effective radiation to the audience area

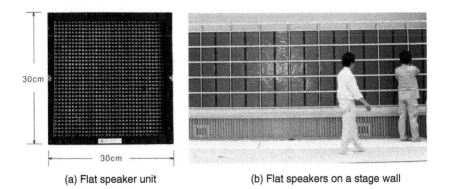

(a) Flat speaker unit   (b) Flat speakers on a stage wall

Figure 10.13 An example of the flat, large plane loudspeaker installed in a lecture hall. Reproduced by permission of Y. Yamasaki (2010).

is obtained without harmful sound radiation to the ceiling, which would increase reverberant sound. Consequently, reverberation is depressed and high intelligibility may be obtained. Furthermore, the sound divergence is in two dimensions. It means sound reduction by distance is nearly 3dB/DD (Figure 5.2). Therefore, sound distribution in the room approaches a nearly uniform condition.

When a flat, wide loudspeaker is used, as shown in Figure 10.13, there is scarcely any divergence, so that the sound reduction with distance is very small and sound distribution becomes almost uniform. Further notable merit of this flat distribution is the near prevention of howling, since the microphone is not exposed to a strong sound.

e. Distribution system for sound field control

Sound field control includes localisation of the sound images and the adding of reflections and reverberation, etc.

1   *Control of localisation of the sound images*: In order to localise the sound images at the positions of the original sound sources, the arrival time and sound level of each signal from each loudspeaker should be controlled at each listening point. The situation with a single original source on the stage has been covered in the above. If there are many original sources on a wide stage, a **delta stereophony system** is applicable, though it is complicated. The basic principle is shown in Figure 10.14. The stage sound is received by i distributed microphones and fed to j distributed loudspeakers through an $(i \times j)$ matrix circuit, consisting of delays and level controllers, with which the sound images are controlled, so that they are localised at the microphone positions. In order to amplify the direct sounds i of the j loudspeakers are situated close to the microphones. Also in this case, it is important to realise the indispensable fundamentals, (1) of Section b, illustrated in Figure 10.2. When the ceiling is too high, the loudspeakers may be better installed in/under chairs or on the floor.

2   *Enhancement of sound reflection and reverberation*: In order to improve the sound field in a room, the early reflected sound is sometimes reinforced by an electro-acoustic device at the seats, where the early sound reaches only weakly, as shown in Figure 10.15. This device is called an **active reflector**, since the sound pressure received by a microphone situated on the wall surface is amplified and emitted from a loudspeaker on the wall. By combining many such devices, prolongation of the reverberation time has been attempted. The most difficult problem is to suppress howling, and many systems for resolving it have been developed. The oldest successful system is in the Royal Festival Hall in London (1964), with more than 150 channels with microphones, each limited within a very narrow frequency band by a Helmholtz resonator.

Normal modes of room vibration are enhanced, hence the system is called **assisted resonance**. Later, systems, which have wider frequency bands, and which combine the electronic reverb-machine or a signal-processing device for adding many sound reflections, have appeared. All of them have many distributed loudspeakers and their layout plays a very important role (Berkhout 1988).

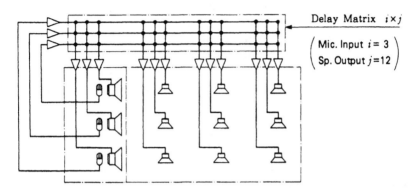

Figure 10.14 Basic principle of delta stereophony for good localisation of the distributed sound source (Steinke 1983).

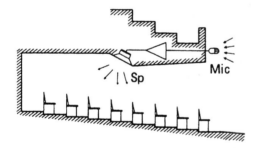

Figure 10.15 Reinforcing a sound reflection by an electro-acoustic system.

However, the most important thing is to have an electronic device that is not noticed by the audience, since almost all classical music fans do not like electronic sound. Nowadays, because of improved technology, most famous opera houses and theatres successfully use electronic acoustic devices, employing a skilled audio-master.

10.3 Electro-acoustic equipment

The instruments that are frequently used are listed in Table 10.1, according to their purpose. Basically, the received sound signals from microphones and other input signals are fed into the input system, after mixing and amplifying to the proper level, then sent to the output system, which supplies enough power to the loudspeakers for sound radiation. An outline of the equipment is described below.

A. Basic construction of reinforcement systems

A diagram of a simple reinforcement system is shown in Figure 10.16: (a) is a block diagram of the circuit of the system; and (b) shows the level diagram of its performance. The reference value of 0 dB is the voltage of 0.775 V, which supplies an electric power of 1 mW = 1/1000 W to a circuit of 600 Ω impedance. Generally, since line input levels from recorded signals, etc., are higher than microphones, a step attenuator and phase selector are set up (Step I in Figure 10.16) and a head amplifier (HA) amplifies the signal. The frequency equaliser (EQ) (Step II) controls the sound quality of the signal, and the attenuator or fader (Vol) controls the sound level. The boost amplifier or buffer amplifier (BA) (Step III) amplifies the signal to the standard input level of the power amplifier (PA). The PA sends the signal to the loudspeaker (Sp) with sufficient electric power to drive it and radiate the sound, with impedance matching between the output of the PA and the input of the loudspeaker. Actually, Steps I, II and III are constructed compactly as one **input module** (Figure 10.17) and a number of input modules for many microphones and other auxiliary channels are placed in a row to create a **mixing console**. Here, all input signals are selected, mixed and controlled by a skilful operator, as shown in the next section.

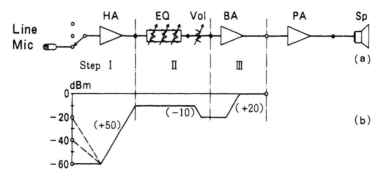

Figure 10.16 (a) Block diagram of a basic circuit; and (b) level diagram of an input module of a sound reinforcement system.

B. Sound mixing and control equipment

Through a matrix circuit, as shown in Figure 10.18, which receives a number of input channels *i* and mixes them into *j* group channels, the sound

Figure 10.17 Example of an input module of a sound mixing console.

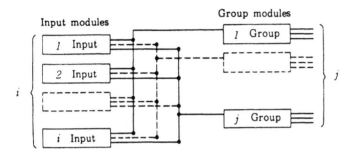

Figure 10.18 Matrix circuit for grouping *j* channels from *i* input channels.

quality and volume of the mixed signals are controlled with **group modules** similar to the input module. The output signal from a group module is fed to a power amplifier, which drives a loudspeaker system. Recently, since the number of channels both of input and output has increased, by using a $(j \times k)$ matrix circuit, *j* output signals are often rearranged into *k* **program modules** for connecting the outputs to the power amplifiers.

The associated **delay system** or **reverb-machine**, etc., including the **compressor/limiter**, which maintains the sound signal at its proper volume, are called **effects** or **effecter**. It generally receives the signal from a group module and feeds back to the same group module after processing the signal. The attenuator of the group module is called a **master fader** and plays an essential role in sound control. In addition, the mixing console has a **talkback module** for audience announcement, a **phones module** which, with headphones, enables an arbitrary channel to be selected and monitored, and also **VU meters** or **peak level meters**, which indicate the working level of every channel.

Generally, a mixing console does not have a monitor system, but is manufactured so that it controls a sound field where the console is situated, or the sound field produced by a monitor system designed for a special purpose, such as music recording, broadcasting or performance of popular or modern rock music, etc. Therefore, when a mixing console is installed in a control room separated from the auditorium, an adequate monitor system must be consciously designed as described previously, in order to monitor the sound fields in the auditorium, and also to satisfy the other requirements.

A **graphic equaliser** (**GEQ**) is used to comprehensively control the transmission frequency characteristics of each loudspeaker channel, including the sound field in the room, by changing the gain of each 1/3 octave band, placed in close proximity to each other's channel. And, also, the console will have **parametric equalisers** (**PEQ**) with a narrower band, and sometimes a **notch filter** at a set frequency for the suppression of howling.

As mentioned above, the number of operating switches and the variety of control knobs is very large, since one or more of 20–30 modules has several tens of each arranged in a line on an operating panel of the control console. Therefore, a mixing console should be designed for ease of operation, simplified and automated, so that it is well within the capabilities of the operator. It is desirable that, as far as possible, and in order so that a control is not touched by mistake, a transparent cover is used for most of the control switches and knobs, except those which are frequently used, such as faders. Moreover, there should be properly illuminated spots on the graphically-displayed circuit to prevent mistakes in operation and to make the system user-friendly. Recently, these facilities have been made possible through digital technology, and are easy to perform using a computer screen. Also, automatic control facilities are available, so that the automatic progress of a performance can be controlled with the aid of computer memory according to a preset program.

C. Equipment for input and output systems

a. Microphone

The construction of three popular types of microphone are illustrated in Figures 10.19–10.21, and their features are shown in Table 10.2.

A single stereo condenser microphone is also available for music and drama. So far, dynamic microphones, and more recently, electret condenser microphones, are the most popular and are inexpensive. However, both their price and characteristics are so wide-ranging that it is important to obtain expert opinion in order to make a sensible choice.

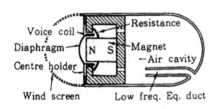

Figure 10.19 Section of a dynamic microphone.

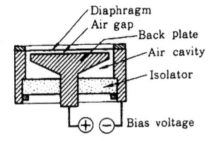

Figure 10.20 Section of a condenser microphone.

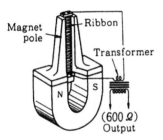

Figure 10.21 Construction of a velocity microphone.

b. Wireless microphone

Wireless microphones are widely used because they are easily moved. The special receiver must be equipped for receiving VHF or UHF, which is transmitted from the microphone. General problems are tone quality and

Table 10.2 Microphones: types and features

Types and directional characteristics	Features
Dynamic microphones (moving coil type), unidirectional and omnidirectional	Strong construction, easy operation with stable performance; various styles for many purposes; also varied ability of performance.
Condenser microphones, unidirectional and omnidirectional (also variable)	Flat frequency characteristics; needs a bias voltage and installation of a head amplifier; high output level; sensitive to moisture; simple usage for electret type with a dry cell; very popular because of small size and low price.
Velocity microphone (ribbon type), bidirectional (figure of eight)	Favourite for sound quality of human voice and music; large in size and weight; low output level; low frequency enhancement as microphone approached. Easily affected by air flow, studio use only.

stability of performance. Since dead spots occur due to electromagnetic standing waves in the room, the **diversity receiving system**, which automatically selects the strongest signal using double antennae, is recommended. Since, again, the quality and price vary widely, any choice needs to be made with great care.

c. Recording and playback devices

Analogue 45–33 1/3 rpm discs, **LPs,** have already been superseded by digital recordings on compact discs (**CDs**), though some people have continued to love the LP's tone quality. And now, many more kinds of digital recording system are in development, such as SACD and DVD-audio, aiming for higher quality.

Regarding magnetic tape recorders with analogue technology, **compactcassette** recorders have been most popular and widely used, and the open-reel tape recorder has also been employed professionally for a long time because of its reliability and convenience for editing and cueing. Nowadays, a number of digital audio devices have been developed. Initially, digital audio tape recorders (**DAT**) were much in favour because of their high quality, now, unfortunately, they are difficult to find on the market. Digital recordable discs (**CD-R**) and recordable-erasable discs (**CD-RW**) are now in general use due to the popularity of personal computers. Digital discs have many superior features, including a large dynamic range and SN ratio, and long-life stability with non-contact pickups, etc.

Regarding recording media, many kinds of solid memory using semiconductors are now being developed rapidly, with IC technology, and they have excellent features, i.e. with no moving parts, so there are never any mechanical troubles when recording and playing back. The development of digital audio technology will be continued endlessly.

Therefore, when selecting equipment, it is preferable to follow the current trend, though with care, in order to have compatibility with common media. So, all electro-sound systems should always be capable of accepting the latest equipment as a part of the overall sound system.

d. Loudspeakers

The loudspeaker has been mentioned already from the user's point of view. Here, its mechanism and features will be explained in more detail.

1   Mechanical construction of loudspeakers

Currently, cone- and horn-type loudspeakers, shown in Figures 10.22 and 10.23, are widely used. Both are electrodynamic. The cone loudspeaker radiates directly from a diaphragm made of paper or plastic. Its efficiency, however, is so low that the acoustic power output is about 1% of the electrical input power. The horn loudspeaker has a small diaphragm, made from

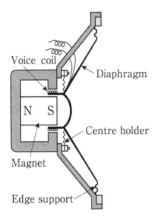

Figure 10.22 Section of a cone loudspeaker.

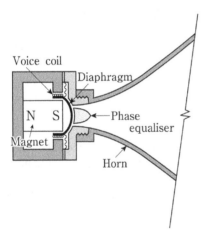

Figure 10.23 Section of a horn loudspeaker.

plastic or light metal in a narrow space, called a **compression driver**, and a long horn is attached in front of it. The horn significantly increases efficiency by more than a factor of 10, as its cross-section area gradually increases from the small area at the diaphragm to a large area at the open mouth of the horn. It works as an acoustic transformer, with high impedance at the compression driver and low impedance at the open mouth. The electrical impedance of both types of loudspeaker is as low as a few to tens of ohms.

There is another type of loudspeaker called a **dome tweeter**. It has a hemi-spherical dome as a diaphragm, as shown in Figure 10.22, but without the surrounding larger diaphragm, and is used only for a low- or mid-power loudspeaker system (Figure 10.11), since it has very low efficiency.

The smaller loudspeakers of a distributed system usually comprise a single- or double-cone of 10–30 cm diameter, driven by a high-impedance constant voltage line through transformers, and cover the full frequency range, 100–10 000 Hz, with electrical power of 10–100 W. As for Figure 10.2, this system is applicable for each of five channels having different delay times.

The larger loudspeakers on a stage or of a centralised system (Figure 10.7) usually consist of a multi-way arrangement, i.e. a cone loudspeaker of 30–45 cm diameter in a bass-reflex cabinet, or combined with a short horn for low frequencies up to 200–1000 Hz, and one or more horn loudspeakers for mid- and higher-frequency ranges. They are directly driven by low-impedance lines from the power amplifier output, where the nominal power of the program input is more than several hundred watts.

The control of the directivity of a horn loudspeaker is achieved by means of the shape of the throat and horn, as shown at the top of Figure 10.7. Recently, several complete sets of 2 or 3-way loudspeaker boxes are often assembled to provide the required directivity, as shown in Figure 10.11.

The tone quality of a cone loudspeaker is subjectively pleasing, with a natural soft sound, especially for the human voice. However, when greater power is needed in a large space, e.g. for the reinforcement of modern music, the horn loudspeaker has become indispensable because of its high efficiency.

2   Outside use

For outdoors public address systems, a folded-horn loudspeaker, called a **trumpet-speaker**, is frequently used, because it is weatherproof, and has high efficiency. However, though it has a frequency range of about 200–5000 Hz, which is enough for speech, there are often problems due to sound quality, caused by erratic and poor operation.

When there is a musical concert in the open area, a huge number of big loudspeaker systems, including woofers with high power, are operated. It is important to consider the problem of noise nuisance caused by the high sound levels.

e. Power amplifier

A power amplifier should have an output with sufficient electrical power and appropriate impedance in the audio frequency range to match the input of the loudspeaker system. The maximum power output should have sufficient margin so as not to distort peak signals and have a protection circuit to deal with extravagant inputs. Furthermore, the SN ratio, dynamic range, percentage distortion and damping factor (load impedance/internal impedance), etc. are all part of the technical specification. The distortion should be small, but the larger the other parameters the better.

Power output has grown over the years, and now several hundred watts are usually available. But, it is desirable to drive a complex system with many loudspeaker units with a multi-amplifier system, so that each speaker unit is directly driven by an independent power amplifier with a low impedance line, in order to obtain high-quality sound. There is another recent trend, that, for convenience, a speaker system carries its own power amplifier, thus needing only a signal cable. It is also easy to move to anywhere where there is an existing AC power supply.

The total AC power consumption is two to three times the total nominal power of the amplifiers, including a safety factor for the peak load. Further, considering the upward trend in consumption, a cooling fan is indispensable, and do not forget to control the fan noise and to deal with the heat load on the air-conditioning system in the amplifier room.

10.4 Prevention of howling

Howling should not occur under any conditions. In the feedback loop in Figure 10.24, howling will occur when the loop gain exceeds unity, i.e. when the sound pressure received by the microphone from the loudspeaker is greater than the initial sound pressure received by it. Therefore, the loop gain should be kept as low as possible in order to obtain a stable operation.

A. Reduction of direct sound

The direct sound level L_d from the loudspeaker (Sp) to the microphone (Mic), shown in Figure 10.24, should be reduced. In order to satisfy this:

1 the speaker and microphone should be separated from each other, paying attention to the directional perception of the audience;
2 a barrier such as a wing wall is useful;
3 the microphone should be outside the area covered by the loudspeaker's directivity;
4 a unidirectional microphone should be used.

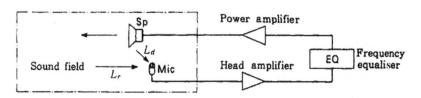

Figure 10.24 Feedback loop of a sound reinforcement system.

B. Receiving a high signal level

The direct sound of the original source should be received at a high sound pressure level. For this purpose, the microphone should be as close as possible to the source. For instance, it is very useful to attach a small microphone to the speaker's lapel.

C. Reduction of reflected and diffused sound

The reflected and diffused sound level L_r at the microphone, shown in Figure 10.24, should be reduced. For this purpose:

1   the radiated sound from the loudspeaker should be controlled and directed to the audience area or other absorptive surfaces, and not to the reflective walls or ceiling;
2   the total absorption in the room should be increased, resulting in a shorter reverberation time;
3   the surfaces surrounding the microphone should be absorptive, especially the surface covered by the microphone's directivity, i.e. the stage wall should be treated with sound absorbent.

D. Flattening of frequency characteristics

The total frequency response of the feedback loop should be flat. If there is a peak, oscillations will occur at that frequency. Therefore, all instruments, such as microphones and loudspeakers, etc., should be of high grade and with flat frequency responses. Standing waves in the room should be eliminated using sound diffusion or absorptive treatment. Furthermore, a frequency equaliser (EQ) as in Figure 10.16, and a GEQ and a PEQ or a **notch filter**, described in Section 10.3.B, are especially useful for eliminating the peak frequency and obtaining the flat frequency characteristics of the total circuit.

10.5   Architectural planning and design

A. Layout planning of a sound control room

All electro-acoustic equipment, except the distributed loudspeakers, are installed in a sound control room in which the whole system is operated in such a way as to give full control of all of its functions. The basic guidelines for planning this room are as follows:

1   there should be an extensive view of the whole audience and the stage, and there should be good contact with the stage;
2   it should be possible to monitor the sound in the audience area.

In Figure 10.4, four positions of the sound control room are shown. Position A in the figure satisfies condition (1) but the distance to the stage is too large. Position B is best, though the balcony seats are out of view. Position C is preferred, especially in public halls, because it is easy to manage the stage preparation with a small number of staff through close contact with the stage. About 10–15 m² of floor area is needed for a mixing console and accompanying instruments and, additionally, 5–10 m² for an annex.

The best position for the window is in front of the operator's seat and it should be of sufficient size to give a good view of the stage floor and suspended microphones, and it should have an openable sash with curtain. The interior must be treated with an absorbent, both for sound and light, in order to avoid leakage to the audience area, and the room should have an independent air-conditioning system. Positions D_1 and D_2 in Figure 10.4 are ideal for satisfying condition (2), since these are representative of what the audience receives, though several of the best audience seats are missing.

According to the character and use of the hall, and after studying its system of operation, the control room should be positioned at A, B or C. Positions D_1 or D_2 are also possible for auxiliary use.

B. Monitoring system in the control room

All control work, generally, should have a perfect real-time monitoring system, which gives an instant response to a control operation in order to feed back to the next operation. However, a monitoring system that is conscious of the feedback loop is rare. Apart from positions D_1 or D_2 inside the auditorium, a monitoring system that gives virtual reality for the operator in a separate and closed room is the most important instrument, playing the role of a nerve centre. Its indispensable functions are as follows:

1   the sound pressure level (dBA) at a representative audience seat should be indicated on the main part of the mixing console;
2   the sound at the representative audience seat should be heard with a binaural or stereo system.

The level indicator is required to cover a wide range, at least from 50–100 dBA, with no switching, and a binaural listening system is desirable for virtual reality. The position of the monitor microphone is critical. If its position is poor, it works only as a noise monitor or so-called air-monitor. A suitable position for it must be found by hanging it from a catwalk on the ceiling, and close to the stage in the directional zone of the main loudspeaker system in the centre of the proscenium arch. The best position would be slightly off the centre line of the hall, in order not to interfere with the spotlight to the stage or offend the eyes of the audience.

Ideally, the system should be able to monitor all sound, not only that from the loudspeakers, and indicate to the operator the sound pressure level in the audience area. For this purpose, the directivity of the monitor microphone is useful.

C. Annex rooms

It is desirable that the electro-acoustic hardware, such as amplifiers, etc., is installed in a separate room or enclosed by a glass screen for observation, so as to isolate the cooling-fan noise. It is also a good idea to enclose the control cubicle with a glass wall with a good view from the control console. The storeroom for microphones, cables, microphone stands, recording media (discs or tapes, etc.), depending on the scale of the equipment, should be located close to the control room. It is important that the room containing the microphones and recording tapes, etc. are air-conditioned as they can be damaged by dust and moisture.

D. Wiring and microphone arrangement

The microphone elevator on the stage should be planned in cooperation with the building engineer, so that the mechanism does not foul any structural component such as a beam or column, and there should be a clearance of 2.6 m below the stage floor to allow for a hydraulic jack. A three-point hanging microphone has three winches attached to the ceiling, so that it can be easily moved to any position within a triangular area limited by the three locating points of the winches. It should be free from any air stream caused by air-conditioning, and from stage lighting.

There should be many microphone sockets set in the stage, floor, wainscot, ceiling and in the audience area for performances that require amplification of sounds there. After collecting the microphone inputs at a socket panel at the side of the stage, they are fed to an input board in the control room. Their cables should be laid out so they do not run close to power lines and loudspeaker cables and are shielded from stray electromagnetic fields in order to avoid induced signals, particularly from the lighting controller.

10.6   Performance test of electro-acoustic systems

The performance of a first class electro-acoustic system should satisfy the following requirements: (a) high intelligibility of speech; (2) rich and high quality of music sound; and (3) safe handling without howling. After the hearing test, including various kinds of short samples of music and speech, several physical measurements are carried out as follows.

A. Measurement of maximum sound pressure level

Pink noise, in the frequency range 63–8000 Hz, is played in place of music, or 125–4000 Hz in place of speech, from the main loudspeaker system. The long-term stability of the maximum sound pressure level (10–30 min) at the centre of the audience area should be measured and should satisfy the recommended values in Table 10.3.

Table 10.3 Recommended maximum sound pressure levels with electro-acoustic systems (dB)

Lecture room	75–80
Conference room or banquet room	80–85
Sports facilities in open field	85–90
Multi-purpose auditorium (popular music)	90–95
Disco (rock music)	105–110

Average values: equipment must not distort at a peak of more than 10 dB higher.

B. Transmission frequency characteristics

The measurement described in Chapter 3, Section 3.6.B, is carried out at the centre of the audience area.

The measured values should be compared with the recommended ranges in Figure 10.25. For a monitor loudspeaker system in a small studio or listening room, flat characteristics are desirable. In a large auditorium, however, it would sound so unnatural that the frequency characteristics would be better controlled to produce natural sounding music and speech, and improved intelligibility obtained by using a graphic equaliser.

In order to check intelligibility, as an alternative to a direct measurement of the **percentage articulation,** as described in Chapter 1, Section 1.14.A, measurements of **STI** or **RASTI** (see Section 3.6.E) are more common.

C. Sound pressure distribution in steady state

The measurement described in Section 3.6.A is carried out in the room. The uniform distribution in the whole audience area might be ideal, but it is acceptable that measured values fall in the range of 6 dB or so. Irregularities of the distribution are liable to be caused by standing waves at low frequencies and by the directivities of loudspeakers at high frequencies.

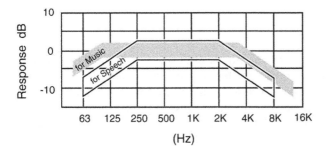

Figure 10.25 Recommended transmission frequency characteristics in auditoria.

Therefore, it makes sense to control the directivity of loudspeakers at high frequencies.

D. Safe amplifying gain for reinforcement

In the circuit shown in Figure 10.24, when the loop gain is set at -6 dB from the howling point, which is called the **howling margin**, the sound pressure level at a representative seat of the audience area, referred to the sound pressure level at the microphone, is called the **safe amplifying gain**. The higher this value, the better, but should generally equal -10 dB at least. Though it varies with the directivity of the microphone, etc., the sound level of -10 dB is equivalent to an attenuation with distance of a factor of 3. Therefore, when a man speaks at 30 cm from the microphone, the listener in the seat can hear with a sound level equivalent to a distance of 90 cm from the speaker.

This measurement is undertaken after the completion of all other measurements and controls mentioned above. As shown in Figure 10.26, in place of the original speaker, a small loudspeaker radiates pink noise at a distance of 30–50 cm from the stage microphone, at two different sound pressure levels and measured at the stage microphone position and at the audience seat.

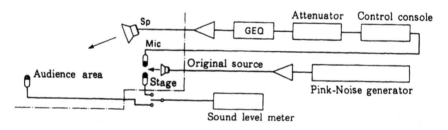

Figure 10.26 Block diagram for measuring the safe amplifying gain of a sound reinforcement system.

E. Remaining noises

Noises can be caused by the power supply, i.e. a hum at low frequencies and random noise from circuit elements at high frequencies. The noise level caused by the loudspeaker system should be lower than the threshold of hearing (Chapter 2, Figure 2.6), while the system is in operation. The highest noise level must be 10 dB lower than the recommended values shown in Table 2.5.

PROBLEMS 10

1 When planning the equipment of an electro-acoustic system in an auditorium, list those items that the architect needs to consider.
2 In planning a public address system for a room, compare the merits and demerits of the centralised system and the distribution system of loudspeakers, and describe how to organise both systems.
3 List all procedures necessary to control howling from a sound reinforcement system.
4 'Difficult to make it howl' is an important ability of a reinforcement system. What should be measured in order to evaluate this?
5 When a mixing console is installed in a control room, separated from an auditorium, what consideration needs to be given to the monitoring system?

11 Addenda

11.1 Wave equations (Lit. B7, B13, B20, B32, etc.)

A. Bulk modulus

The changes in pressure when a sound wave travels through air are, in general, so rapid that heat cannot be exchanged between different volume elements. Consequently, the changes are adiabatic. In air, when the initial pressure and volume are P_0 and V_0, respectively, are changed to $(P_0 + p)$ and $(V_0 + \Delta V)$, respectively, due to sound pressure p, the following relationship is obtained

$$P_0 V_0^\gamma = (P_0 + p)(V_0 + \Delta V)^\gamma$$

$$\therefore 1 + \frac{p}{P_0} = \left(1 + \frac{\Delta V}{V_0}\right)^{-\gamma}$$

where γ is the ratio of specific heats at constant pressure and constant volume. If $\Delta V / V_0$ is very small, then expanding the right-hand side of the above equation and approximating, it follows that

$$\frac{p}{P_0} = -\gamma \frac{\Delta V}{V_0}$$

$$\therefore p = -\gamma P_0 \frac{\Delta V}{V_0}$$

Putting

$$\kappa = \gamma P_0 \tag{11.1}$$

$$p = -\kappa \frac{\Delta V}{V_0} \tag{11.2}$$

Therefore, sound pressure is proportional to the volume change. κ is called the **bulk modulus of elasticity** or **volume elasticity**.

B. *Plane wave equation*

Consider a tube of unit cross-sectional area with its axis parallel to the direction of propagation of a plane wave, as shown in Figure 11.1, where the plane at x is displaced by ξ and the plane at $(x + \delta x)$ is displaced by

$$\xi + \delta\xi = \left(\xi + \left(\frac{\partial \xi}{\partial x}\right)\delta x\right)$$

This means that the original volume $V = \delta x$ contained between the two planes has been increased by

$$\Delta V = \left(\frac{\partial \xi}{\partial x}\right)\delta x$$

due to the passage of the sound wave. Thus, the fractional increase in volume is

$$\frac{\Delta V}{V} = \frac{\partial \xi}{\partial x}$$

which is the so-called equation of continuity. Substituting this into Equation (11.2) gives

$$p = -\kappa \frac{\partial \xi}{\partial x} \tag{11.3}$$

When the sound pressure p acts on the plane at x, then

$$\left(p + \left(\frac{\partial p}{\partial x}\right)\delta x\right)$$

acts on the plane at $(x + \delta x)$. The differential pressure

$$\left(\frac{\partial p}{\partial x}\right)\delta x$$

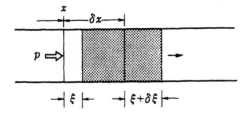

Figure 11.1 Plane wave propagation.

sets in motion the mass of air between the two planes. If ρ is the average density of the air, then the mass between the two planes is $(\rho\delta x)$. Hence, using Newton's second law, which states that force equals mass times acceleration, we have

$$
\rho\delta x\frac{\partial^2\xi}{\partial t^2} = -\frac{\partial p}{\partial x}\cdot\delta x
$$
$$
\therefore\ \rho\frac{\partial^2\xi}{\partial t^2} = -\frac{\partial p}{\partial x}
$$
(11.4)

This is the equation of motion.

Differentiating both sides of Equation (11.3) with respect to t and Equation (11.4) with respect to x and eliminating ξ,

or

$$
\frac{\partial^2 p}{\partial t^2} = \frac{\kappa\partial^2 p}{\rho\partial x^2}
$$
(11.5a)

$$
\frac{\partial^2 p}{\partial t^2} = c^2\frac{\partial^2 p}{\partial x^2}
$$
(11.5b)

where

$$
c=\sqrt{\frac{\kappa}{\rho}}
$$

This is the wave equation in terms of sound pressure. Similarly, eliminating p, the wave equation, in terms of the displacement ξ is obtained in the same form. Then, introducing a new function, the velocity potential φ defined by Equation (3.2) in Chapter 3,

$$
\left.\begin{array}{l}
\dfrac{\partial\xi}{\partial t} = -\dfrac{\partial\varphi}{\partial x}\\[2mm]
p=\rho\dfrac{\partial\varphi}{\partial t}
\end{array}\right\}
$$
(11.6)

and using the fundamental Equation (11.3), the following is obtained

$$
\frac{\partial^2\varphi}{\partial t^2} = c^2\frac{\partial^2\varphi}{\partial x^2}
$$
(11.7)

This has exactly the same form as Equation (11.5) for sound pressure.

If the sound wave behaves as a simple harmonic vibration, then the function may be written as $\varphi\, e^{j\omega t}$ where ω is the angular frequency, then Equation (11.7) may be written

$$
\left(\frac{d^2}{dx^2}+k^2\right)\varphi=0
$$
(11.8)

where

$$k = \frac{\omega}{c}$$

k is called the **wavelength constant** or **wave number**.

C. *Characteristic impedance of media*

The solution of Equation (11.8) can be expressed in the form

$$\varphi = C_1 e^{j(\omega t - kr)} + C_2 e^{j(\omega t + kr)}$$

where C_1 and C_2 are constants. The first term describes a wave propagating in the positive x direction, while the second term describes a wave propagating in the negative x direction.

When there is only one wave propagating in one direction, that is, a progressive plane wave, $C_2 = 0$; therefore

$$\varphi = C_1 e^{j(\omega t - kx)} \tag{11.9}$$

So from Equation (11.6)

$$v = \frac{\partial \xi}{\partial t} = jkC_1 e^{j(\omega t - kx)}$$
$$p = \rho j \omega C_1 e^{j(\omega t - kx)}$$

Thus, the impedance is a real number only,

$$Z = \frac{p}{v} = \rho c \tag{11.10}$$

This is called the **characteristic impedance** of the medium, but it must be noted that this is for a plane wave in a homogeneous medium of infinite extent.

D. *Wave equation in three-dimensional space*

In order to derive the wave equation for the propagation of a plane wave in three-dimensional space, we use the same principles as for the one-dimensional case, applying the equations of continuity and motion of the medium. Then it follows

$$\frac{\partial^2 \varphi}{\partial t^2} = c^2 \left(\frac{\partial^2 \varphi}{\partial x^2} + \frac{\partial^2 \varphi}{\partial y^2} + \frac{\partial^2 \varphi}{\partial z^2} \right) \tag{11.11}$$

Equation (11.11) can be written

$$\frac{\partial^2 \varphi}{\partial t^2} = c^2 \nabla^2 \varphi \tag{11.12}$$

When the sound wave is a simple harmonic motion, since φ can be replaced by $\varphi\, e^{j\omega t}$ the above equation may be written

$$\left(\nabla^2 + k^2\right)\varphi = 0 \tag{11.13}$$

where $k = \omega/c$.

E. Spherical wave equation and acoustic impedance

In order to derive a wave equation, we transform to a spherical coordinate system with polar coordinates (r, θ, ψ) in which the wave function does not depend on θ or ψ but on r and t. Hence the wave equation may be written

$$\frac{\partial^2(\varphi r)}{\partial t^2} = c^2 \frac{\partial^2(\varphi r)}{\partial r^2} \tag{11.14}$$

A general solution of this equation, which is finite everywhere except at $r = 0$, is

$$\varphi = \frac{1}{r} F(ct - r) + \frac{1}{r} G(ct + r)$$

In free space, since a sound wave is radiated outward from the sound source, the first term alone exists, and, in the case of simple harmonic motion, the above expression reduces to

$$\varphi = \frac{A}{r} e^{j(\omega t - kr)}$$

From Equation (11.6) the following are obtained

$$v = -\frac{\partial \varphi}{\partial r} = \left(\frac{1 + jkr}{r}\right)\frac{A}{r} e^{j(\omega t - kr)} \tag{11.15a}$$

$$p = \rho \frac{\partial \varphi}{\partial t} = j\omega\rho \frac{A}{r} e^{j(\omega t - kr)} \tag{11.15b}$$

Therefore, the acoustic impedance density is

$$Z = \frac{p}{v} = \frac{j\omega\rho r}{1 + jkr} = \rho c \frac{jkr}{1 + jkr} \tag{11.16}$$

When the distance r is very large $(kr >> 1)$, the impedance density approaches ρc, which is equivalent to the case of a plane wave, although it must be noted that, in the near field of the sound source, this situation does not hold true. Also, the sound pressure, i.e. Equation (11.15b), is inversely proportional to r, while the particle velocity given by Equation (11.15a) is inversely proportional to r^2 in the near field with $kr << 1$. This is the reason why low-frequency sounds are unnaturally emphasised when a velocity microphone is brought too close to any sound source, e.g. the mouth, often a problem in practice.

11.2 Analogy between electrical, mechanical and acoustic systems and the time constant (Lit. B13, B40, etc.)

A. Correspondence between electrical and mechanical systems

When an electromotive force $E(t)$ is applied to an inductance, as shown in Figure 11.2, an electric current i is generated according to Faraday's law as follows

$$L\frac{di}{dt} = E(t) \tag{11.17}$$

The electric charge q in this circuit is related to the current by

$$i = \frac{dq}{dt} \tag{11.18}$$

therefore

$$L\frac{d^2q}{dt^2} = E(t) \tag{11.19}$$

Also, when $E(t)$ is applied to a capacitor C, as shown in Figure 11.3, the generated electric charge q will be given by

$$\frac{q}{C} = E(t) \tag{11.20}$$

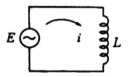

Figure 11.2 Electric circuit with an inductance.

Figure 11.3 Electric circuit with a capacitor.

and the relation to the electric current i is

$$\frac{1}{C} \int i \, dt = E(t) \tag{11.21}$$

On the other hand, when an external force $F(t)$ acts on a mass m, a velocity v results (see Figure 11.4). According to Newton's second law, mass, velocity and force are related by

$$m \frac{dv}{dt} = F(t) \tag{11.22}$$

Expressing this in terms of the displacement ξ,

$$m \frac{d^2 \xi}{dt^2} = F(t) \tag{11.23}$$

Also, when $F(t)$ acts on a spring whose elastic modulus is k, producing a displacement ξ at the end, as shown in Figure 11.5

$$k\xi = F(t) \tag{11.24}$$

Figure 11.4 A force acting on a mass.

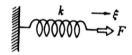

Figure 11.5 A force acting on a spring.

Table 11.1 Analogy between electrical and mechanical systems

Electrical system	Mechanical system
Inductance, L	Mass, m
Current, i	Velocity, v
Charge, q	Displacement, ξ
Electromotive force, $E(t)$	External force, $F(t)$
Reciprocal of electrostatic capacitance, $1/C$	Elastic modulus, k

If v is the velocity at the tip then

$$k \int v dt = F(t) \tag{11.25}$$

Now comparing Equations (11.17)–(11.19) and (11.22)–(11.23) with Equations (11.20)–(11.21) and (11.24)–(11.25), respectively, and the corresponding terms shown in Table 11.1, it can be seen that the equations for both electrical and mechanical systems have exactly the same form. Therefore, if the solutions of Equations (11.17)–(11.19) and (11.20)–(11.21) are found, those of Equations (11.22)–(11.23) and (11.24)–(11.25) are also known. In this case, these two systems are said to be equivalent and, furthermore, a problem in one system can be solved by analogy with the other. Thus, mechanical or acoustic vibrating systems can often be solved easily using alternating current theory if they are replaced by an equivalent electrical circuit, since circuit theory has been extremely well developed.

B. Simple resonant system

The equation of motion of a single resonance system with a mass on a spring, as discussed in Chapter 7, Section 7.4, is

$$m\frac{d^2\xi}{dt^2} + r\frac{d\xi}{dt} + k\xi = F(t) \tag{11.26}$$

Rewriting the above in terms of velocity v,

$$m\frac{dv}{dt} + rv + k \int v dt = F(t) \tag{11.27}$$

The system is illustrated in Figure 11.6(a).

It can be seen by comparing Equations (11.17) and (11.22) that, in the first term of the above equation, the mass m is equivalent to induction and, by comparing Equations (11.21) and (11.25), the modulus of elasticity k in the third term is equivalent to the reciprocal of capacitance, that velocity v corresponds to the current i and, hence, frictional resistance in a mechanical

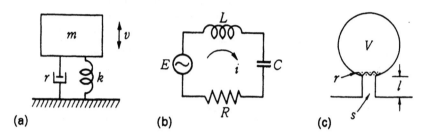

Figure 11.6 Single resonance system: (a) mechanical system; (b) electrical system; and (c) acoustic system.

system corresponds to electrical resistance in an electrical system. Therefore the following equation, corresponding to Equation (11.27) is obtained

$$L\frac{di}{dt} + Ri + \frac{1}{C}\int idt = E(t) \tag{11.28}$$

So the equivalent circuit is a well-known L-R-C circuit as shown in Figure 11.6(b). If the electromotive force E is a simple harmonic alternation then $E(t) = E\,e^{j\omega t}$ where $\omega = 2\pi f$. Remembering $d/dt = j\omega$ Equation (11.28) can be written

$$i\left(j\omega L + R + \frac{1}{j\omega C}\right) = E e^{j\omega t} \tag{11.29}$$

Hence, the solution is

$$i = \frac{E e^{j\omega t}}{R + j(\omega L - (1/\omega C))} \tag{11.30}$$

where

$$Z_e = R + j\left(\omega L - \frac{1}{\omega C}\right) \tag{11.31}$$

is the electrical impedance of this circuit, and, at the frequency at which the imaginary part $(\omega L - 1/\omega C)$ becomes zero, resonance occurs. Namely, when

$$\omega L = \frac{1}{\omega C} \qquad \therefore \ \omega^2 = \frac{1}{LC}$$

The resonance frequency is therefore

$$f_r = \frac{1}{2\pi}\sqrt{\frac{1}{LC}} \tag{11.32}$$

The mechanical impedance of the mechanical system of Figure 11.6(a) is by analogy

$$Z_m = r + j\left(\omega m - \frac{k}{\omega}\right) \tag{11.33}$$

And, similarly, setting the imaginary part equal to zero, the resonance frequency is

$$f_r = \frac{1}{2\pi}\sqrt{\frac{k}{m}} \tag{11.34}$$

The vibration velocity is

$$v = \frac{1}{Z_m} \cdot Fe^{j\omega t} \tag{11.35}$$

The displacement is

$$\xi = \frac{v}{j\omega} = \frac{1}{j\omega Z_m} \cdot Fe^{j\omega t} \tag{11.36}$$

and the acceleration is obtained as follows

$$\frac{dv}{dt} = j\omega v = \frac{j\omega}{Z_m} \cdot Fe^{j\omega t} \tag{11.37}$$

C. Frequency characteristics

We now consider how the characteristics of the mechanical impedance $Z_m = j\omega m + r + k/j\omega$ may change when subject to a wide range of frequency variation (see Figure 6.15).

(i) When the frequency is lower than the resonance frequency

$$f = \frac{1}{2\pi}\sqrt{\frac{k}{m}}$$

the first term for Z_m is much smaller than the third term, so that over a certain range of frequencies

$$Z_m \cong \frac{k}{j\omega} \tag{11.38}$$

In this range, since the vibration is determined by the elastic modulus k, the vibration is referred to as stiffness-controlled under which condition the displacement is constant by Equation (11.36).

(ii) Near the resonance frequency, the imaginary part, that is, the sum of the first and third terms, approaches zero; hence

$$Z_m \cong r \tag{11.39}$$

which is independent of frequency. In this frequency range, since displacement, velocity and acceleration amplitude are determined by resistance, the vibration is said to be resistance-controlled, and, as can be seen from Equation (11.35), the velocity is constant.

(iii) In the range where the frequency is much higher than the resonance frequency, the first term is quite large, then the impedance may be approximated as follows:

$$Z_m \cong j\omega m \tag{11.40}$$

This condition is then referred to as inertia-controlled or mass-controlled, and from Equation (11.37) the acceleration dv/dt is constant.

D. Acoustic vibration system

As discussed in Chapter 4, Section 4.5, the Helmholtz resonator, as shown in Figure 11.6(c), corresponds to the mechanical and electrical systems illustrated in Figures 11.6(a) and 11.6(b).

Consider a single degree of freedom resonance system in which the mass m of air in the neck vibrates on the air cushion in the cavity, which acts like a spring, where the trapped air has a bulk modulus κ. Equations (11.26) and (11.27) apply to this system. In the resonator the mass m of the air in the neck is given by

$$m = sl\rho \tag{11.41}$$

where s is the cross-sectional area of the neck, l the effective length and ρ the density of air.

In order to determine the elastic modulus of the volume V of air in the cavity, we first assume a piston whose area is s and which is subjected to an inward displacement x. So the trapped air is compressed due to a pressure increase

$$p = \kappa \frac{sx}{V}$$

Then, from Equation (11.2) of Section (11.1), the force F applied to the piston is given by

$$F = sp = \kappa \frac{s^2 x}{V}$$

but the elastic modulus

$$k = \frac{F}{x}$$

$$\therefore k = \frac{\kappa s^2}{V} \qquad (11.42)$$

Furthermore, due to the resistance r per unit area of the neck, the viscous loss is $s\,r\,v$. On the other hand, since the external force acting on the resonator is due to an alternating sound pressure $p\ e^{j\omega t}$ per unit area, the following equation is obtained

$$sl\rho \frac{dv}{dt} + srv + \frac{s^2 \kappa}{V} \int v\,dt = spe^{j\omega t} \qquad (11.43)$$

From this, comparing with Equation (11.33), the following expression

$$Z_m = \frac{sp}{v} = sr + j \left(\omega sl\rho - \frac{s^2 \kappa}{\omega V} \right) \qquad (11.44)$$

is the mechanical impedance of the resonator.

Since the acoustic impedance Z_a is defined as the ratio of sound pressure to volume velocity

$$Z_a = \frac{p}{sv} = \frac{Z_m}{s^2} \qquad (11.45)$$

It follows therefore that the specific acoustic impedance Z_{sp} is

$$Z_{sp} = \frac{p}{v} = sZ_a = \frac{Z_m}{s} \qquad (11.46)$$

The resonant frequency of this resonator is thus

$$\omega^2 = \frac{s\kappa}{l\rho V}$$

Since the imaginary part of any impedance is zero at resonance, from Equation (11.5) of Section 11.1

$$\kappa = \rho c^2 \quad \therefore \; \omega^2 = \frac{sc^2}{lV}$$

$$\therefore \; f_r = \frac{c}{2\pi}\sqrt{\frac{s}{lV}} \tag{11.47}$$

When dealing with more complicated acoustic vibration systems than this example, one can appreciate that such an analogous solution, based on the corresponding mechanical or electrical systems, is much more effective.

E. Time constant of an integrating circuit

(a) Let us consider the transient phenomenon when a direct current voltage E is applied to the input of a series RC circuit as shown in Figure 11.7. From Figure 11.3 and Equations (11.18), (11.20)

$$R\frac{dq}{dt} + \frac{1}{C}q = E \tag{11.48}$$

The solution of this equation is

$$q = CE + Ke^{-t/RC} \tag{11.49}$$

This is called a delay system of the first order, which means the exponential term is of the first order. Since the integration constant K is determined by the initial condition, putting $q = 0$ at $t = 0$, it follows that $K = -CE$

$$\therefore \quad q = CE\left(1 - e^{-t/RC}\right) \tag{11.50}$$

The output voltage is

$$e_0 = \frac{q}{C} = E(1 - e^{-t/RC}) \tag{11.51}$$

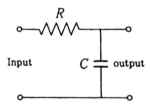

Figure 11.7 Electrical *RC* series circuit.

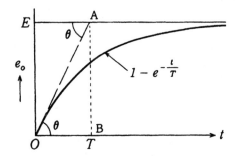

Figure 11.8 The output of an *RC* circuit as a function of time.

where

$$RC = T \qquad (11.52)$$

the so-called time constant.

The output voltage varies, as shown in Figure 11.8, $e_0 \to E$ as $t \to \infty$. If a line tangent to the curve at $t = 0$ is drawn, it intercepts $e_0 = E$ at angle θ. The point of interception is A and the corresponding point on the abscissa is B, as shown in the figure. Differentiating Equation (11.51)

$$\tan \theta = \frac{de_0}{dt} = \frac{E}{RC} = \frac{AB}{OB} \quad \therefore \quad OB = RC = T$$

i.e. 63.2% of the steady state value and when $t = T$, the output voltage becomes

$$e_0 = E(1 - e^{-1}) = 0.632E \qquad (11.53)$$

(b) When a fluctuating voltage $E(t)$ is applied to the circuit shown in Figure 11.7, from Equation (11.48)

$$\frac{dq}{dt} + \frac{q}{T} = \frac{1}{R}E(t) \qquad (11.54)$$

When T is sufficiently large, the second term on the left-hand side of the equation can be neglected; hence

$$q \cong \frac{1}{R} \int E(t)dt$$

$$\therefore \quad e_0 = \frac{q}{C} \cong \frac{1}{T} \int E(t)dt \qquad (11.55)$$

The output voltage is proportional to the integral of the input voltage, therefore the *RC* circuit shown in Figure 11.7 is called an integrating circuit.

(c) The indicator of a sound level meter or a vibration-level meter consists of a squaring circuit and an *RC*-integrating circuit, in order to give the root-mean-square value of the signal and the time-weighting characteristics, as defined by the time constant of the integrating circuit. Therefore, the peak value of the input of the toneburst signal is as shown in Figure 11.9, which is a transformed presentation in dB of the curve in Figure 11.8. When the duration of the toneburst signal is the same as the time constant of the circuit, the peak value is 2 dB below the steady-state value according to Equation (11.53).

(d) In the frequency domain, since the impedance of this circuit is $Z = R + 1/j\omega C$, at the frequency where $R = 1/\omega C$, $\omega = 1/RC = 1/T$, and the output power is 1/2 the input power and the circuit functions as a low-pass filter with a cut off frequency

$$f_c = \frac{1}{2\pi \cdot T} \tag{11.56}$$

as shown in Figure 11.10.

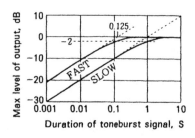

Figure 11.9 Output of *RC* circuit with different time constants.

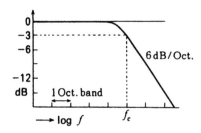

Figure 11.10 The frequency response of an *RC* circuit.

11.3 Fourier transformation and the correlation function (Lit. B39, B40, etc.)

The basic components of signal processing that are frequently applied in acoustic measurements are formulated here, but with the omission of mathematical proofs.

A. Fourier transform

Generally, a time function $f(t)$ can be expressed using the Fourier integral as follows,

$$f(t) = \frac{1}{2\pi} \int_{-\infty}^{\infty} F(\omega)e^{j\omega t} d\omega \tag{11.57}$$

where the function $F(\omega)$ is given by

$$F(\omega) = \int_{-\infty}^{\infty} f(t)e^{-j\omega t} dt \tag{11.58}$$

This is called the Fourier transform of $f(t)$, and $f(t)$ is called the inverse Fourier transform of $F(\omega)$. Both are complex functions of amplitude and phase, and have perfectly reciprocal relations.

The coefficient $1/2\pi$ in Equation (11.57) can be removed and shifted to Equation (11.58), or $1/\sqrt{2\pi}$ can be used as a coefficient for both Equations, (11.57) and (11.58), in other words, the product of the two coefficients should be $1/2\pi$.

In physical terms, the Fourier transform $F(\omega)$ of a wave form $f(t)$ in the time domain is a frequency spectrum. Figure 11.11 shows representative wave forms and their frequency spectra. The absolute value of $F(\omega)$ is called the amplitude spectrum and its square the power spectrum.

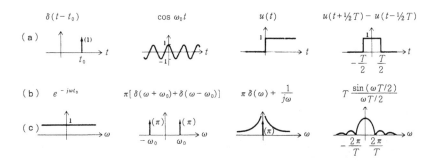

Figure 11.11 Examples of wave forms and spectrums: (a) wave form $f(t)$; (b) Fourier transform $F(\omega)$ and (c) amplitude spectrum.

B. *Unit impulse function and convolution integral*

A unit impulse function is called a delta (δ) function and is defined mathematically as

$$\delta(t - t_0) = 0 \quad t \neq t_0 \tag{11.59a}$$

also

$$\int_{-\infty}^{\infty} \delta(t - t_0)\, dt = 1 \tag{11.59b}$$

δ is a time function, having the integral value of unit area, but an amplitude that is zero everywhere except at $t = t_0$, where the amplitude is infinite.

When the unit impulse $\delta(t)$ is fed to a linear transmission system S, the output signal is called the impulse response of the system. Then, if a signal $x(t)$ is fed to this system, the output is expressed by

$$y(t) = \int_{-\infty}^{\infty} x(\tau)h(t - \tau)\, d\tau = \int_{-\infty}^{\infty} x(t - \tau)h(\tau)\, d\tau = x(t) * h(t) \tag{11.60}$$

This is called a convolution integral and is often represented by an asterisk ($*$). Figure 11.12 gives a visual explanation of this, applied to the principle of superposition in a linear system. This relation plays a very important and widely applicable role in various linear systems.

When the impulse response is expressed in the form of a Fourier integral, a pair of Fourier transforms is obtained as follows

$$h(t) = \frac{1}{2\pi} \int_{-\infty}^{\infty} H(\omega)e^{j\omega t}\, d\omega \tag{11.61}$$

$$H(\omega) = \int_{-\infty}^{\infty} h(t)e^{-j\omega t}\, dt \tag{11.62}$$

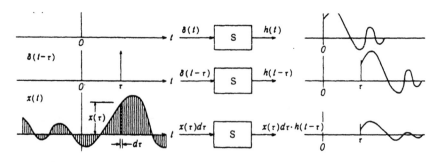

Figure 11.12 Explanation of convolution integral with the superposition theorem.

where $H(\omega)$ is called a system function or transfer function.

If the Fourier transform $X(\omega)$ of the input signal $x(t)$ is known, then the Fourier transform $Y(\omega)$ of the output $y(t)$ is

$$Y(\omega) = X(\omega) \cdot H(\omega) \tag{11.63}$$

This means that the output in the frequency domain is obtained simply by multiplying the Fourier transform of the input signal by the transfer function.

C. *Autocorrelation function*

In order to express the properties of a time-variant wave form, the autocorrelation function, as defined by Equation (11.64) is used

$$\varphi_{11}(\tau) = \int_{-\infty}^{\infty} f_1(t) f_1(t+\tau) \, dt \tag{11.64}$$

especially for an irregular function,

$$\varphi_{11}(\tau) = \lim_{T \to \infty} \frac{1}{2T} \int_{-T}^{T} f_1(t) f_1(t+\tau) \, dt \tag{11.65}$$

It may be sufficient to integrate over one period if the function is periodic, and for an aperiodic function, such as music or speech, an integration time of several or several tens of milliseconds may be sufficient. Figure 11.13 shows examples of the autocorrelation functions of simple periodic functions.

Since the value of $\varphi_{11}(\tau)$ is a maximum at $\tau = 0$, the normalised form with maximum value

$$\phi_{11}(\tau) = \frac{\varphi_{11}(\tau)}{\varphi_{11}(0)} \tag{11.66}$$

is often used. Figure 11.14 shows examples of the autocorrelation function for noise, music and speech displayed using Equation (11.66).

The autocorrelation function $\varphi_{11}(\tau)$ and the power spectrum $\Phi_{11}(\omega)$ make a pair of Fourier transforms as follows

$$\varphi_{11}(\tau) = \int_{-\infty}^{\infty} \Phi_{11}(\omega) e^{j\omega t} \, d\omega \tag{11.67}$$

$$\Phi_{11}(\omega) = \frac{1}{2\pi} \int_{-\infty}^{\infty} \varphi_{11}(\tau) e^{-j\omega t} \, d\tau \tag{11.68}$$

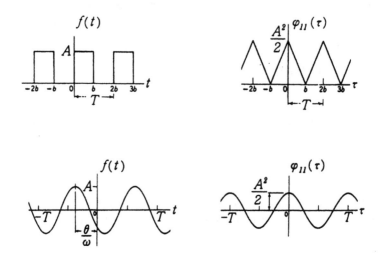

Figure 11.13 Examples of autocorrelation functions of periodic waves.

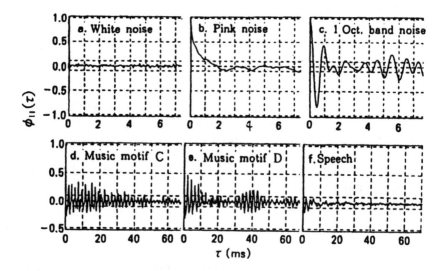

Figure 11.14 Examples of autocorrelation functions of aperiodic signals.

When $\tau = 0$ the autocorrelation function becomes an integral of the square of $f_1(t)$

$$\varphi_{11}(0) = \int_{-\infty}^{\infty} f_1^2(t)\mathrm{d}t \tag{11.69}$$

This gives the total energy of $f_1(t)$

D. Cross-correlation function

In order to express the similarity of two time functions $f_1(t)$ and $f_2(t)$ the cross-correlation function $\varphi_{12}(\tau)$, generally defined by Equation (11.70), is frequently used.

$$\varphi_{12}(\tau) = \lim_{T \to \infty} \frac{1}{2T} \int_{-T}^{T} f_1(t)f_2(t+\tau)dt \qquad (11.70)$$

In practice, the normalised form of the above, that is, Equation (11.71) is often used

$$\phi_{12}(\tau) = \frac{\varphi_{12}(\tau)}{\sqrt{\varphi_{11}(0)\varphi_{22}(0)}} \qquad (11.71)$$

The results of psychoacoustic research on room acoustic assessment, show that the less the degree of similarly between signals $f_l(t)$, and $f_r(t)$, which are received by the left and right ears, respectively, that is, the lower the value of $\varphi_{12}(\tau)$, then the larger the spatial impression. Ando (see Lit. B33) defined the IACC (Inter-Aural Cross-Correlation) as follows,

$$\text{IACC} = |\varphi_{lr}(\tau)|_{\max}, \qquad |\tau| \leq 1 \text{ ms} \qquad (11.72)$$

This is used as a parameter for the evaluation of sound fields. Figure 11.15 shows an example of this.

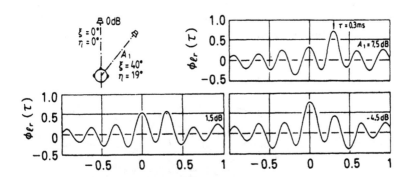

Figure 11.15 Example of calculated results of IACC when the intensity of a single reflection is changed (Y. Ando 1985, Lit. B33).

11.4 Outline of auditory organ (Lit. B5, B28, B34, Chapter 4, etc.)

The structure of the human ear is shown in Figure 11.16. Sound collected by the pinna (auricle lobe) passes through the outer ear canal and oscillates the eardrum, which is at the entrance to the middle ear. The oscillation of

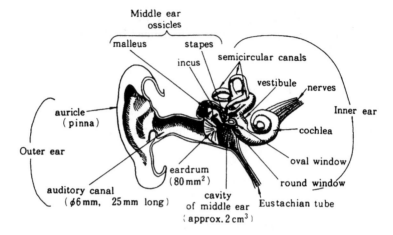

Figure 11.16 Cross-section through human ear.

the eardrum is transferred by three bones called ossicles in the middle ear to the oval window, which closes the entrance to the inner ear. Three ossicles, the malleus (hammer), incus (anvil) and stapes (stirrup), which are of different lengths, create a lever ratio (impedance transformer) that provides impedance matching between the air oscillation over a large area and the motion over a small area of lymph (a fluid) in the cochlea, which comprises the inner ear.

The cochlea is coiled like a snail shell whose cross-section consists of three scalae (canals) as shown in Figure 11.17. The extended length of it is about 35 mm while the scala vestibuli and the scala tympani are connected

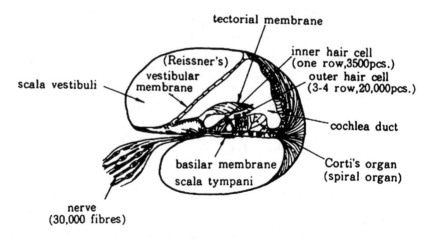

Figure 11.17 Cross-section through the cochlea.

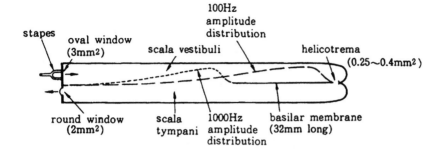

Figure 11.18 Schematic representation of unrolled cochlea.

to the helicotrema, a small window at the apex of the cochlea as shown in Figure 11.18. The scala media (or cochlear duct) is separated from the other two scalae by Reissner's membrane and the basilar membrane and is filled with lymph, as shown in Figure 11.17. The basilar membrane is loaded with the organ of Corti (or Corti's organ), which has sensory hair cells embedded at both sides of the trigonal pillar. The hairs of both inner and outer hair cells are in touch with the tectorial membrane.

When the oval window receives an oscillation, transferred from the middle ear, the travelling wave produces a vertical displacement, which occurs in the basilar membrane with the lymph in the scala media; the hairs are bent by the relative displacement of hair cells and the tectorial membrane. Then, spikes (electric pulses) are fired electro-physiologically in the auditory nerves, which transmit the spikes to the brain.

The amplitude of displacement of the basilar membrane is small near the oval window, but the position of maximum displacement depends on the frequency, as shown in Figure 11.18, where the higher the frequency the closer to the oval window is the maximum displacement. It is considered that, to some extent, a frequency analysis is performed. Although the variation of electric potential produced by the hair cells is related to the sound wave, as if the cochlea plays a similar role to that of a microphone, the electric pulses transmitted to the brain are of quite a different style. Therefore, the generation mechanism, the transmission system in the auditory nerves and the faculty of auditory sensation in the cerebral cortex are researched under the umbrella of electrophysiology.

11.5   Calculation of loudness level (A) by E. Zwicker's method

Though human sensory perception cannot be directly measured, several methods have been proposed to measure the physical magnitude of a stimulus. Here and in the following Section, 11.6, two methods standardised by ISO 532 (1975) are explained.

The subjective magnitude of sound or noise is expressed as the loudness level L_N with the unit phon or the loudness N with the unit sone, see Figure 1.12. While the loudness level is a logarithmic measure (like the dB scale) and the increase of the level by 1 phon is approximately the differential limen for a subjective sensation, the loudness is a measure directly proportional to the subjective magnitude of the sound strength. The relation between loudness N and loudness level L_N is

$$N = 2^{(L_N-40)/10} \qquad\qquad (11.73)$$

E. Zwicker developed a method for calculating the loudness level of sound that has been analysed in terms of 1/3 octave bands, which was first published in 1960. This method offers some advantages in relation to applications in building acoustics and environmental acoustics, where the use of 1/3 octave-band filters for measurements is well established. The result of the method is either the loudness level for frontal sound in phons (GF) if the listener is in the open air, or the loudness level for a diffuse field in phons (GD) if sound reaches the listener from all directions as in an ordinary room.

The purpose is to provide a method by which complex sounds of various levels and spectra may be ordered on a scale of subjective magnitude. The method is applicable not only to sounds with broadband spectra but also to sounds with tones or irregular spectra. The method is based on the concept of critical bands. A critical band is defined as the widest frequency band in which the loudness level depends only on the sound pressure level. The audible frequency range is divided into 24 critical bands (with the unit Bark). Above about 300 Hz the critical bands are approximately the same as 1/3 octave bands, but at lower frequencies the critical bands are wider.

Originally, Zwicker's method was a graphical method; the 1/3 octave band levels are transformed into areas that correspond to specific loudness in critical bands. The loudness level is calculated from the total area for the entire audible frequency range. However, with the development of PCs, the method has also been transferred to computer programs (Zwicker *et al.* 1991).

Since frequency perception in human hearing is best described by critical bands, the first step in the method is to approximate the critical bands by the usual 1/3 octave bands. At lower frequencies the critical bands are wider than the 1/3 octave bands, and thus the following approximations are applied for the first three critical bands:

L_1: The total sound pressure level of 1/3 octave bands with centre frequency 80 Hz and below (energy summation as in Equation (1.21)),

L_2: The total sound pressure level of 1/3 octave bands with centre frequencies 100, 125 and 160 Hz,

L_3: The total sound pressure level of 1/3 octave bands with centre frequencies 200 and 250 Hz.

Table 11.2 Example of 1/3 octave band spectrum transferred to critical band levels

Frequency (Hz)	L_p (dB)	Critical band (bark)	L_i (dB)
50	65		
63	65	1	69.5
80	64		
100	64		
125	61	2	66.2
160	56		
200	54	3	57.0
250	54		
315	54	4	54.0
400	51	5	51.0
500	50	6	50.0
630	44	7	44.0
800	41	8	41.0
1000	39	9	39.0
1250	38	10	38.0
1600	38	11	38.0
2000	44	12	44.0
2500	49	13	49.0
3150	49	14	49.0

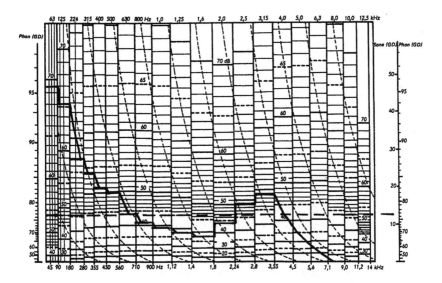

Figure 11.19 Graph for estimation of the loudness level according to Zwicker's method (ISO 532). The spectrum given in Table 11.2 is shown as an example. This figure is reproduced with the permission of the Danish Standards Foundation. The figure is based on the Danish Standard DS/ISO 532:1975. © Danish Standards Foundation.

In the 1/3 octave bands with centre frequencies 315 Hz and higher, the sound pressure levels are used directly as approximations to the critical band levels. An example is shown in Table 11.2.

Next, these levels L_i are drawn as horizontal lines covering the actual frequency band in the graph, with stepped curves from ISO 532, see Figure 11.19. Where the steps are rising with frequency, the adjacent horizontal levels are connected by vertical lines, but when the level in the next frequency band is lower, a downward sloping curve is drawn as an interpolation between the dashed curves on the graph. The area under this curve corresponds to the total loudness. If this area is transformed into a rectangle with the same area, the loudness level can be read from the scale on either side of the graph. The spectrum given in Table 11.2 is drawn into Figure 11.19 and the result in this example is 76.5 phon (GD).

11.6   Calculation of loudness level (B) by Stevens's method

From 1940 onwards S. S. Stevens pursued his research on this subject and published a series of reports from Mark I (1956) to Mark VII (1972). The procedure is as follows: first, find the loudness index S_i (sone) from measured values, in either 1, 1/2 or 1/3 octave bands of the objective noise, using Figure 11.20, and obtain the maximum value of S_m. Then the total loudness S_t for the entire frequency range is given by

$$S_t = S_m + F(\Sigma S_i - S_m) \text{ (sone)} \tag{11.74}$$

where

$$\begin{cases} F = 0.3 & \text{(1 Octave band)} \\ F = 0.2 & \text{(1/2 Octave band)} \\ F = 0.15 & \text{(1/3 Octave band)} \end{cases}$$

From the above, the loudness level, L_N (phon) can be calculated by

$$L_N = 40 + 10 \log_2 S_t = 40 + \frac{10}{0.3} \log_{10} S_i \tag{11.75}$$

Also the scale provided at the right-hand side of Figure 11.20 may be used for conversion between sones and phons. Equation (11.74) is a simplified version, taking into account the masking effect due only to the frequency band that has the maximum stimulus. So, this method is much simpler than Zwicker's method, described above in Section 11.5.

Stevens (1972) also published Mark VII as a synthesised evaluation scale, combining not only loudness but also noisiness and annoyance, which is

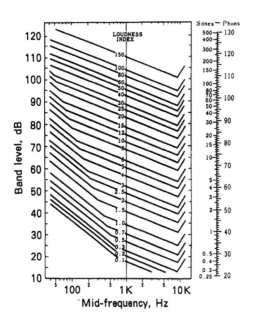

Figure 11.20 Calculation chart for loudness level (Stevens, S.S., Mark VI).

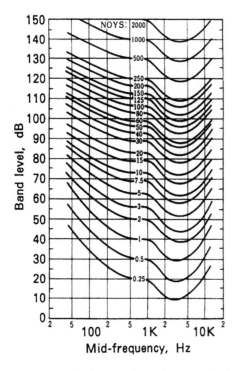

Figure 11.21 Calculation chart for perceived noise level (K.D. Kryter 1970, Lit. B21).

called the perceived level, represented by a chart similar to Figure 11.21 in Section 11.7, not with curves but folded lines, though its applicability requires further investigation. (See Lit. B24.)

11.7 Calculation of noisiness

A. Perceived noise level (PNL)

K. D. Kryter (1970, Lit. B21) realised by experiments on auditory perception that actual noisiness is different from the loudness level (phon) and established a calculation method for **PNL** (Perceived Noise Level) by constructing equal noisiness curves following Stevens's method.

The procedure is exactly the same as in Steven's method, using Equations (11.74) and (11.75) in Section (11.6), except that it employs Figure 11.21 in place of Figure 11.20, and instead of the loudness index S, it uses a value N called **Noy**, hence, the PNL (PNdB) is obtained. The values range from 500 to 1000 Hz in Figure 11.21, corresponding to Equation (11.75) in Section (11.6) (see Lit. B21, B24).

B. Effective PNL (EPNL)

This scale was recommended as ISO-R507:1970 and ICAO Annex 16, for assessment of aircraft noise following the American method and practice, which is based on the total noise power for a fixed-time duration, modified to take into account human sensation and psychology.

Assuming access to computer processing, first a 1/3 octave-band analysis of the fluctuating noise is carried out at 0.5 s intervals, then, after applying pure tone correction due to adjoining band level differences using Figure 11.22, the PNdB is obtained from Figure 11.21. The resultant value is called **PNLT** (tone corrected PNL) and the peak value, when the total energy is approximated, in practice, by the integrated value in the duration D in Equation (11.76), is transformed into an equivalent rectangular shape

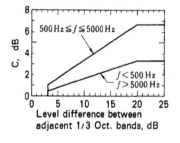

Figure 11.22 Tone correction factors for perceived noise level.

within the prescribed duration time T_o (here 10 s), which is referred to as the **EPNL** (effective PNL). The main purpose of this method is to certify the noise generated by newly manufactured civil aircraft.

The following Equation (11.76) seems, however, appropriate as an approximation for general noise monitoring and land-use planning

$$EPNL = PNL + C + 10 \log_{10} \frac{D}{20} \qquad (11.76)$$

where PNL (PNdB) is obtained from the peak value of the 1 octave-band analysis; C is the pure tone correction, +2 dB, used only for landing of turbofan aircraft; and D is the duration in seconds when the level is 10 dB lower than the peak value. The third term, that is, the duration correction, is based on the assumption of a triangular shape for level change.

11.8   Reverberation in a coupled room

Generally, a theatre has two major spaces, the stage and the seating area, linked via a proscenium opening. Therefore, when calculating the reverberation time of the seating area, the proscenium opening is considered as an absorptive surface, the absorption coefficient of which is assumed. Sometimes, for instance, when the reverberation time is measured at some point during construction, the calculated values never agree with those measured, due to the fact that absorption in the stage space is much lower compared with that in the seating area, as described in Chapter 3, Section 3.3.D. Therefore, we must examine the situation where two spaces exchange acoustic energy mutually through the opening between them. Such a combined space is called a coupled room for which Eyring (1931) developed a solution. The basic principle of dealing with this problem is simply explained as follows.

As shown in Figure 11.23, a room R_1 is coupled with another R_2 through the opening F. We then assume that (1) both rooms have completely diffuse sound fields and (2) the acoustic energy density changes abruptly at the opening between the rooms.

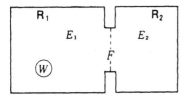

Figure 11.23 Coupled room.

When the source located in room R_1 generates acoustic power W, the energy density is E_1 and E_2 in the two rooms, respectively, and the surface areas, except for the area of the opening, are S_1 and S_2, with average absorption coefficients α_1 and α_2, respectively. Hence, the energy absorbed in room R_1 is $cE_1(S_1\alpha_1 + F)/4$, whereas the energy returned from room R_2 is $cE_2F/4$ (see Section 3.3.B). Therefore, the total energy in room R_1 changes according to the following equation

$$V_1 \frac{dE_1}{dt} = W - \frac{cE_1(S_1\alpha_1 + F)}{4} + \frac{cE_2F}{4} \tag{11.77}$$

Similarly, in room R_2

$$V_2 \frac{dE_2}{dt} = -\frac{cE_2(S_2\alpha_2 + F)}{4} + \frac{cE_1F}{4} \tag{11.78}$$

These two equations should be solved simultaneously.

In the steady state

$$\frac{dE_1}{dt} = \frac{dE_2}{dt} = 0$$

While, during decay, $W = 0$. Hence the form of the solution is

$$E_1 = Ae^{-\beta_1 t} + Be^{-\beta_2 t} \tag{11.79}$$
$$E_2 = Ce^{-\beta_1 t} + De^{-\beta_2 t} \tag{11.80}$$

Thus, the problem is reduced to finding values of A, B, C, D, β_1 and β_2 that satisfy Equations (11.79) and (11.80).

Equations (11.79) and (11.80) show that the situation in both rooms can be expressed by the summation of two exponential terms; consequently, the decay curve should be bent unless β_1 and β_2 are equal and, furthermore, the profile of the bent curve is changed according to the magnitudes of A and B or C and D.

11.9   Schroeder method of measuring reverberation time

M. R. Schroeder (1965) presented a new method of measuring the reverberation time on the basis that a simple integration taken over the squared impulse response in a single measurement yields the ensemble average of the decay curves, which, in themselves, are unstable, and where measurements using the conventional method require the taking of the average of a number of measurements.

Assume $n(t)$ to be stationary white noise. Its autocovariance function $\langle n(t_1) \cdot n(t_2) \rangle$ depends only on the time difference $(t_2 - t_1)$, since it is stationary, and is zero everywhere except for $t_1 = t_2$ since it is white noise. Thus, we can write

$$\langle n(t_1) \cdot (t_2) \rangle = N \cdot \delta(t_2 - t_1) \tag{11.81}$$

where the brackets $\langle \rangle$ denote the ensemble average, N is the noise power (per 1 Hz), and $\delta(t_2 - t_1)$ is the delta function, as defined in Section 11.3.B.

When the band noise is switched off at $\tau = 0$ after a steady state has been reached, the signal received at a point is expressed by a convolution integral,

$$S(t) = \int_{(-\infty)}^{0} n(\tau) \cdot h(t - \tau) \, d\tau \tag{11.82}$$

where $h(t)$ is the impulse response from the source to the receiving point, including all related factors. The lower limit $(-\infty)$ in the integral means that sufficient time is needed to build up to the steady state. The decay of sound energy, represented by the square of the received signal, is written as a double integral as follows

$$S^2(t) = \int_{(-\infty)}^{0} d\tau \int_{(-\infty)}^{0} d\theta n(\tau) \cdot n(\theta) \cdot h(t - \tau) \cdot h(t - \theta) \tag{11.83}$$

Applying Equation (11.81) to the ensemble average of Equation (11.83)

$$\langle S^2(t) \rangle = \int_{(-\infty)}^{0} d\tau \int_{(-\infty)}^{0} d\theta N \cdot \delta(\theta - \tau) \cdot h(t - \tau) \cdot h(t - \theta) \tag{11.84}$$

Since $\delta(\theta - \tau)$ vanishes except when $\theta = \tau$, and since the integral becomes unity, integration over θ yields

$$\langle S^2(t) \rangle = N \int_{(-\infty)}^{0} h^2(t - \tau) \, d\tau \tag{11.85}$$

or, using the new integration variable $x = t - \tau$

$$\langle S^2(t) \rangle = N \int_{t}^{(\infty)} h^2(x) dx \tag{11.86}$$

This equation shows that the ensemble average of the squared noise decay, which would require a large number of measurements, may be obtained by only a single measurement of the impulse response. Figure 11.24 shows an example of a measurement using this method.

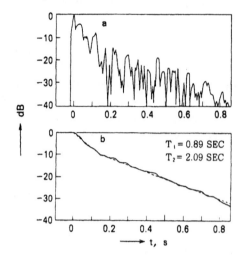

Figure 11.24 Reverberation curves measured in the Philharmonic Hall, New York (Schroeder, M.R.). (a) Toneburst response curve showing sound pressure decay; (b) squared toneburst response integrated from time t to 1 s. T_1 and T_2 show reverberation times obtained by straight-line fits to first 10 dB and remainder of the decay respectively.

11.10   Sound absorption characteristics of multi-layered absorbents

A. Characteristics of sound-absorbing material (Lit. B8, B22)

Since a sound wave is attenuated during transmission through an absorptive material, the sound pressure p of a plane wave propagating in the x-direction is expressed as a function of distance x as follows

$$\left.\begin{array}{l} p = p_0 e^{-\gamma x} \\ \gamma = \alpha + j\beta \end{array}\right\} \tag{11.87}$$

where p_0 is the sound pressure at $x = 0$, γ is the propagation constant, α is the attenuation constant, $\beta = \omega/c$ is the phase constant and c is the sound speed in the material. When $\alpha = 0$, i.e. with no attenuation, β is synonymous with the wave number k.

As regards particle velocity, if the medium is homogeneous and of density ρ, from Equation (11.4) in Section 11.1, $\partial v/\partial t = -1/\rho \cdot \partial p/\partial x$ and in the case of a sinusoidal wave $\partial/\partial t = j\omega$, then

$$v = -\frac{1}{j\omega\rho}\frac{\partial p}{\partial x} \tag{11.88}$$

Assuming the same relationship in an absorptive material, from Equation (11.87) and (11.88)

$$v = \frac{\gamma}{j\omega\rho} p$$

Hence the characteristic impedance of the medium is

$$Z = \frac{p}{v} = \frac{j\omega\rho}{\gamma} \qquad (11.89)$$

Thus, any spatial and time conditions of a sound wave in an infinitely continuous medium can be determined completely by two quantities, i.e. the propagation constant γ and the characteristic impedance Z of the medium.

B. *Single-layer absorbent*

When a single layer of homogeneous material which has thickness l, propagation constant γ and characteristic impedance Z exists with x between 0 and l, as shown in Figure 11.25, and is faced with a surface whose impedance is Z_2 at $x = l$, then what will be the impedance Z_1 at $x = 0$?

If the sound wave propagation in the positive x direction has sound pressure p_i at $x = l$ and the reflected wave returning in the reverse direction has pressure p_r at the same point, the sound wave through the material may be expressed as follows

$$\begin{cases} p(x) = p_i e^{\gamma(l-x)} + p_r e^{-\gamma(l-x)} & (11.90) \\ v(x) = \dfrac{P_i}{Z} e^{\gamma(l-x)} - \dfrac{P_r}{Z} e^{-\gamma(l-x)} & (11.91) \end{cases}$$

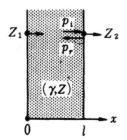

Figure 11.25 Single layer absorbent.

From the boundary condition $p(l)/v(l) = Z_2$, the relation

$$P_r/P_i = (Z_2 - Z)/(Z_2 + Z)$$

is obtained. Substituting this into Equations (11.90) and (11.91), the impedance at $x = 0$ is

$$Z_1 = \frac{p(0)}{v(0)} = Z \cdot \frac{Z_2 \cosh \gamma l + Z \sinh \gamma l}{Z_2 \sinh \gamma l + Z \cosh \gamma l} \tag{11.92}$$

The particular conditions, which apply to the above equation, are as follows:

1   when the material thickness is infinitely large, $Z_2 = Z, \therefore Z_1 = Z$;
2   if the material is fastened to a rigid wall

$$Z_2 = \infty \quad \therefore Z_1 = Z \coth \gamma l \tag{11.93}$$

3   when an air space, whose thickness is $\lambda/4$, is provided between the material and the rigid wall, using $\gamma = jk$ and Equation (11.93) and the assumption that there is no attenuation in air,

$$Z_2 = Z \coth j\frac{\pi}{2} = 0 \quad \therefore \quad Z_1 = Z \tanh \gamma l \tag{11.94}$$

From these relationships the propagation constant γ and characteristic impedance Z of the material can be obtained by measurements that satisfy the requirements of Equations (11.93) and (11.94).

After the acoustic impedance Z_1 of the material surface has been determined from Equation (11.92), the absorption coefficient can be obtained from Equation (1.34).

C. Multi-layered absorbents

In the case of a multi-layered construction, for example, where a composite panel consisting of three kinds of material is directly mounted on a rigid wall, as shown in Figure 11.26, and where each material's propagation

Figure 11.26 Multi-layered absorbent.

constant and characteristic impedance are known, then applying the method described above, firstly the impedance Z_3 can be determined from Equation (11.93) with $Z_4 = \infty$, then substituting Z_3 in Equation (11.92), Z_2 is obtained, and finally Z_1 can be obtained in the same way from Equation (11.92) by substituting Z_2. As a basic principle for any number of layers, the impedance at the surface of the composite wall can be calculated by repeating the above process using Equation (11.92) for the layer that lies at the back. Then the absorption coefficient of the composite wall can be determined from Equation (1.34), although a new calculation method has been published (Mechel 1988).

11.11   Smith chart and standing wave method

A. *Configuration of Smith chart*

Expressing the sound pressure reflection complex coefficient as

$$|r_p|e^{j\Delta} = p + jq \tag{11.95}$$

and from Equations (4.5) and (4.6) the acoustic impedance ratio may be written

$$r + jx = \frac{1 + (p + jq)}{1 - (p + jq)}$$

Therefore, equating real and imaginary parts of both sides, the following relationships are obtained

$$\left(p - \frac{r}{r+1}\right)^2 + q^2 = \frac{1}{(r+1)^2} \tag{11.96}$$

$$(p - 1)^2 + \left(q - \frac{1}{x}\right)^2 = \frac{1}{x^2} \tag{11.97}$$

As shown in Figure 11.27, Equation (11.96) is the locus of constant r for a circle of radius $1/(r+1)$, whose centre has coordinates $p = r/(r+1)$, $q = 0$, while Equation (11.97) is the locus of constant x for a circle of radius $1/x$, whose centre has coordinates $p = 1$, $q = 1/x$.

Since Equation (11.95) yields $|r_p| = \sqrt{p^2 + q^2}$, the locus of constant $|r_p|$ is a circle whose centre is at the origin of the coordinates and where all the values should be within the circle whose radius is 1 because $|r_p| < 1$. The phase angles are shown in Figure 11.28 where $q/p = \tan \Delta$. Figure 11.29 is the Smith chart, which combines Figures 11.27 and 11.28, although concentric circles are omitted and, instead of an angle scale, there is a wave number scale (d/λ) where d is the distance to the first minimum of a standing wave from the specimen surface, as shown in Figure 4.5, and λ is the wavelength.

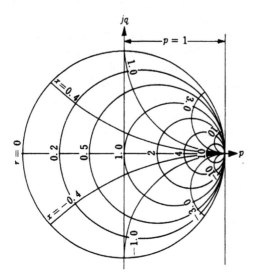

Figure 11.27 Contours of equal sound pressure reflection coefficient in the complex plane.

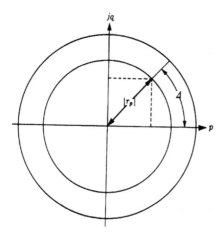

Figure 11.28 The sound pressure reflection coefficient in the complex plane.

B. Using a Smith chart

The Smith chart was intended to be used for calculations on communication transmission lines, providing double scales, one for the wave number towards the load and the other in the opposite direction towards the generator, although in determinations of acoustic impedance derived from

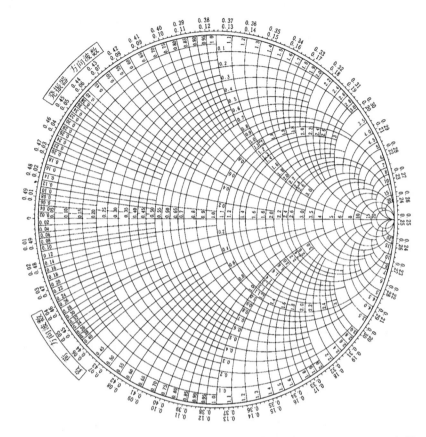

Figure 11.29 Smith chart. Upper half; wavelength towards generator. Lower half; wavelength towards load.

standing waves generated in the tube method, only the outer scale (towards the load) is used.

Wave numbers are graduated from 0 to 0.5 for one rotation starting at the negative end of the real axis in Figure 11.27; then, in the case of a perfect reflection, the first standing wave minimum is at $d/\lambda = 0.25$, which coincides with the positive end of the real axis, thus $\Delta = 0$ in Figure 11.28. In general, d/λ for a material is obtained from the measured value d and then, by connecting the relevant point on the circumference of the circle to the centre, Δ is obtained. Therefore, depending on $d/\lambda \gtrless 0.25$ the phase angle of the reflected wave $\Delta \gtrless 0$ and the imaginary part of the impedance is given by $x \gtrless 0$.

Next, plotting the measured value of the standing-wave ratio n on the real axis, we allow the concentric circle through n to intersect the radius at angle Δ obtained above. The point of intersection gives the acoustic impedance ratio. When the standing wave ratio is measured on the dB scale, the value of n or the radius $|r_p|$ of the concentric circle can be found from Table A.1 in the Appendices. The absorption coefficient can immediately be obtained from the same table.

[**Ex. 11.1**] The measurement of the acoustic impedance of a material is to be carried out using a standing-wave tube. Let us suppose that the room temperature is 15°C and a standing wave is produced in the tube at a frequency of 500 Hz by a tone generator. The results obtained are shown in Figure 11.30 and from Table A.1 it can be seen that:

1   With a standing-wave ratio −13.2 dB (with the minimum that is closest to the sample), $n = 4.57$.
2   The normal-incidence absorption coefficient $\alpha_0 = 58.9\%$.
3   The absolute value of the sound pressure reflection coefficient is

$$|r_p| = 0.64$$

4   The phase angle from Equation (4.4) is given by

$$\Delta = 4\pi\delta/\lambda$$

where λ is the distance from the first to the third minimum, i.e. 824 − 136 = 688 mm. Therefore $\lambda/4 = 172$ mm and so subtracting $d = 136$ mm from this

$$\delta = -36\,\text{mm} \quad (\text{see Figure 4.5})$$

and hence $\Delta = -360° \times 2 \times 36 \div 688 = -37.5°$. Using the Smith chart, $d/\lambda = 136/688 = 0.198$, which is plotted on the wave number scale towards the load and joined to the centre; $\Delta = -37.5°$ is obtained with a protractor. The minus sign indicates a delay of the reflected sound wave at the sample surface.

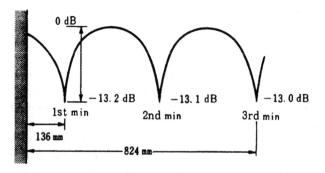

Figure 11.30 Example of a standing wave.

5   To obtain the acoustic impedance from the Smith chart, we draw a concentric circle through $n = 4.57$ on the real axis to intersect a radius at angle Δ, namely the point corresponding to 0.64 of the radius from the centre. For this point $r = 1.5$ and $x = -2.0$, which gives the acoustic impedance ratio at the specimen surface.

11.12   Sound insulation criteria for building in different countries

A. *Airborne sound insulation criteria for walls*

The field measuring method of airborne sound insulation is described in Chapter 6 (see Section 6.2), and the evaluation of the sound insulation of a wall is given by Equations (6.14) or (6.15), i.e. the normalised sound level difference. Following on from this, the ISO single number rating method is described in Section 6.2.D.

B. *Impact sound insulation criteria for floors*

The field measurement of the impact sound insulation of floors is presented as a normalised impact sound level, as described in Chapter 7 (see Section 7.3.B), and the ISO single number rating is derived as described in Section 7.3.C.

C. *Sound transmission class (STC) and impact insulation class (IIC) used in the USA*

STC and IIC are different from the international standards. The single number rating called STC is recommended in the US for rating the airborne sound insulation of walls and is described in ASTM E 413-04. The process of obtaining an STC number is almost the same as ISO 717, described in Section 6.2.D, except that, when the reference curve (called the sound transmission class contour, which is the same as Figure 6.6 but with the frequency range changed to 125 Hz–4 kHz) is shifted towards the measured curve, the maximum unfavourable deviation does not exceed 8 dB at any single frequency.

In order to obtain the rating of the impact sound insulation of a floor, the reference curve is the same as Figure 7.11. However, another scale, IIC, is added, as seen at the right-hand side of Figure 11.31. The reference curve is shifted towards the measured curve of normalised impact sound levels using the same procedure as for STC described above. Figure 11.31 shows three reference curves, which are recommended by the Federal Housing Administration (FHA) for use in different environments, as shown in Table 11.3.

This classification is also applied with the same number to STC for airborne sound insulation. The measured STC and IIC numbers for many

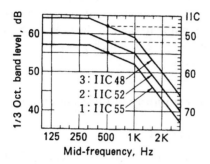

Figure 11.31 IIC contours for classification of impact sound insulation of floors recommended in USA.

Table 11.3 Recommendation for STC and IIC in various environments by FHA in USA

Class	STC & IIC	Environmental condition	Outdoor noise level at night (dBA)
1	55	Rural or high class residential	<40
2	52	Suburban common residential	40–45
3	48	Urban general area	>45

building materials and constructions have been quoted and illustrated in a convenient way for use in noise control (see Lit. A6).

D. Sound insulation criteria in Japan

a. Criteria for airborne sound insulation

In order to evaluate the airborne sound insulation between rooms, the sound level difference, Equation (6.13), is used and the insulation criteria determined, as shown in Figure 11.32 according to JIS (Japanese Industrial Standard) A 1419, and the appropriate classification is as recommended by the Architectural Institute of Japan, as shown in Table 11.4. Building law ensures that the partition wall between dwellings in flats matches the values equivalent to curve D-40.

b. Criteria for impact sound insulation

In Japan, not only the tapping machine but also heavy-weight impact sources are standardised (JIS A 1418) as a simulation of a child jumping. The standard is based on the principle that the jump is equivalent to the impact of an automobile tyre, having a weight 7.3 ± 0.2 kg with an air pressure of

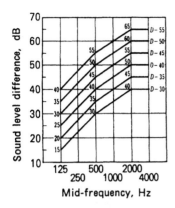

Figure 11.32 Classification of airborne sound insulation with level difference between rooms in Japan.

Table 11.4 Recommendation for average sound level difference between rooms by AIJ (Architectural Institute of Japan) in Japan (1979)

Building	Rooms	Parts	Class S better	1 normal	2 acceptable	3 minimum
Apartment houses	Living	Party wall and floor between dwellings	D-55	D-50	D-45	D-40
Hotel	Bedroom	Party wall and floor	D-50	D-45	D-40	D-35
Office	Office meeting room	Party wall	–	D-40	D-35	D-30
	Needing more privacy	Party wall	D-50	D-45	D-40	–
School	Classroom Lecture room	Party wall	D-45 D-50	D-40 D-45	D-35 D-40	D-30 D-35

$(2.4 \pm 0.2) \times 10^5$ Pa, falling from a height of 0.85 m on the floor but bouncing once only. Now, another standard of a rubber ball, which has about 1/3 impact force, 0.185 m diameter and 2.5 ± 0.2 kg weight, has been added in order not to break the testing floor.

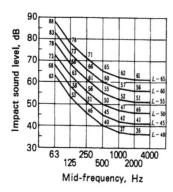

Figure 11.33 Classification of impact sound insulation of floors in Japan.

Table 11.5 Recommendation for impact sound insulation of floor by AIJ in Japan (1979).

Building	Rooms	Building component	Class S	1	2	3
Apartments	Living	Party floor	L-40	L-45	L-50, 55	L-60
Hotel	Bedroom	Party floor	L-40	L-45	L-50	L-55
School	Classroom	Party floor	L-50	L-55	L-60	L-65

A sound level meter and 1 octave-band analyser are used as sound receiver and for measuring the peak values of each single impact with a time weighting of F. The measured curve is then evaluated with the aid of the reference curves in Figure 11.33, which are an upside-down version of the A weighted curves, which simulate human hearing. Shown in Table 11.5 is the appropriate classification based on building uses and locations, as recommended by the Architectural Institute of Japan.

Discussions are taking place so that these standard methods can be recognised internationally.

11.13   Measurement of power level and directivity of sound source

As discussed in Chapter 5 (see Section 5.1), it is convenient to express the sound source output in terms of the power level in order to calculate the sound level at a receiving point. Now, let us consider how to obtain the sound power level in principle. For further detail the reader should refer to ISO 3740.

A. Sound pressure measurement in a free field

Using Equations (5.2), (5.6) and (5.11) the power level L_W can be calculated on the assumption that the sound source is omnidirectional and that the reflected sound has no effect on the field, as would be the case in an anechoic chamber.

When the sound source is directional and the intensity at a point of distance r in the direction θ, φ is $I_{\theta\varphi}$, as shown in Figure 11.34, the total energy radiated for the whole solid angle is given by the following integral

$$W = \int I_{\theta\varphi} r^2 d\omega \tag{11.98}$$

where $d\omega$ is an infinitesimal unit angle. In practice, this may be approximated by the following expression where the solid angle is divided into small portions $\Delta\omega$

$$W = \sum I_{\theta\varphi} \Delta\omega \tag{11.99}$$

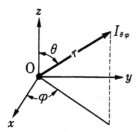

Figure 11.34 Coordinates showing directivity of a sound source.

If the measured sound pressure level at a distance r is $L_{\theta\varphi}$ the intensity $I_{\theta\varphi}$ is obtained from Equation (1.18).

$$I_{\theta\varphi} = \log_{10}^{-1} \frac{L_{\theta\varphi}}{10} \times 10^{-12} \tag{11.100}$$

Substituting this into Equation (11.99),

$$L_W = 10 \log_{10} \left(\sum \log_{10}^{-1} \frac{L_{\theta\varphi}}{10} \Delta\omega \right) + 20 \log_{10} r \text{ (dB)} \tag{11.101}$$

[Ex. 11.2] When a sound source has an axisymmetrical directivity, then locating the measured point in the plane including the axis of symmetry (for example, the z-axis), the individually assigned solid angle $\Delta\omega$ may be obtained from

$$\Delta\omega = \int_{\theta_1}^{\theta_2} 2\pi \sin\theta \, d\theta$$

For example, if every measuring point is taken at 30° spacing, ±15° is to be assigned at the boundaries for each division; thus $\Delta\omega$ for each measuring point will be given by Table 11.6.

Table 11.6 Axisymmetrical case

Measuring point θ	$\Delta\omega$
0°	0.214
30°	1.63
60°	2.82
90°	3.25
120°	2.82
150°	1.63
180°	0.214

[Ex. 11.3] If a point source is located on the ground, it is possible to make n equal divisions of hemispherical surface 2π; then Equation (11.98) is

$$W = \sum_{}^{n} I_{\theta\varphi} r^2 \frac{2\pi}{n} \quad \therefore \quad \frac{W}{2\pi r^2} = \frac{\sum^{n} I_{\theta\varphi}}{n}$$

Thus, W is simply obtained from the average value of n intensities provided the measuring points follow the specified coordinates indicated in Table 11.7.

Table 11.7 Coordinates of measuring points dividing a hemisphere of unit radius into equal surface areas

Division numbers	No	x	y	z
4	1	0	0.82	
	2	0.82	0	
	3	0	−0.82	0.58
	4	−0.82	0	
6	1	0	0.89	
	2	0.85	0.28	
	3	0.53	−0.72	0.45
	4	−0.53	−0.72	
	5	−0.85	0.28	
	6	0	0	1

B. *Sound power measurement in a reverberation room*

When a sound source of sound power W is situated in a room with a high degree of diffusion, the average energy density E in the steady state in the room is expressed by Equation (3.23); therefore

$$W = \frac{EcA}{4} \tag{11.102}$$

where Ec is the sound intensity from Equation (1.16). Then, by measuring the average sound pressure level $\overline{L}_p$ in the room, the power level is obtained from Equations (1.14) and (1.18) as follows,

$$L_W = 10 \log_{10} \frac{W}{10^{-12}} = \overline{L}_p - 6 + 10 \log_{10} A \tag{11.103}$$

where A can be determined from Equation (3.26) by measuring the reverberation time T.

This method is easily employed since it does not depend on the directivity. However, it is essential that the sound field is, in fact, diffuse.

C. *Method of intensity measurement*

The indirect methods described above are such that the sound power is calculated from the measured sound pressure. When the sound power is measured directly by a sound intensity measuring system then the effect of ambient noise can be cancelled in the steady state. In addition to measurements at fixed points, such as those shown in Tables 11.6 and 11.7, by sweeping the intensity probe over the surface enclosing the sound source, the single-value spatial average intensity is easily obtained. Therefore, it has become possible to measure the sound power emitted by a sound source in a noisy environment. Standardisation of this method is found in ISO 9614-2.

D. *Determination of directivity factor*

When an omnidirectional point sound source is located in free space, the directivity factor $Q = 1$. The ratio of the sound intensity in a particular direction to that of an omnidirectional source of the same total power is defined as the directivity factor $Q_{\theta\varphi}$ in that direction. The measurement of directivity should be performed in an anechoic room or the open air with no reflected sound.

[Ex. 11.4] The sound pressure level is measured as 94 dB at 2 m from a sound source whose sound power level is 105 dB. Let us find the directivity factor in that direction.

If the sound source is omnidirectional, from Equation (5.2)

$$L = 105 - 11 - 20 \log_{10} 2 = 88 \, \text{dB}$$

The required directivity factor Q is obtained from Equation (5.3) as follows.

$$10 \log_{10} Q = 94 - 88 = 6 \, \text{dB}$$

The value expressed in dB is called the directivity gain. Thus, the directivity factor $Q = 4$.

11.14   Sound diffraction around a screen and other barriers

A. *General approximation for sound diffraction*

Consider the application of the Fresnel–Kirchhoff approximate theory of diffraction in optics, to an infinite screen, which exists between a sound source S and receiving point P and which has an open area F as shown in Figure 11.35. The sound pressure U(p) at the receiving point may be expressed as follows, with the condition that the dimensions of the open area are sufficiently small compared to the distances to S and P.

$$U(p) = B \int \int_F e^{jkf(\xi,\eta)} \, \mathrm{d}\xi \, \mathrm{d}\eta \tag{11.104}$$

where B is regarded as a constant even though it depends on the source power, wavelength, distances between source, receiving point and the screen, and other geometrical relations. $k = 2\pi/\lambda$, ξ, η are the coordinates

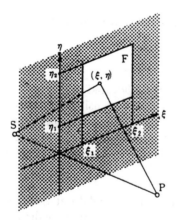

Figure 11.35 Coordinates of an opening in an infinite screen.

in the plane of the screen, and $f(\xi, \eta)$ is a contribution function to the point P. Integration is over the open area F. When the open area is defined by ξ_1, ξ_2, η_1 and η_2, as shown in Figure 11.35,

$$U(p) = B \int_{\xi_1}^{\xi_2} d\xi \int_{\eta_1}^{\eta_2} d\eta . e^{jkf(\xi,\eta)} \tag{11.105}$$

Transforming the variables from $\xi \to u$ and $\eta \to v$, and using an appropriate form of the function $f(\xi, \eta)$ following Kirchhoff's diffraction theory,

$$U(p) = jA \int_{u_1}^{u_2} e^{j(\pi/2)u^2} du \int_{v_1}^{v_2} e^{j(\pi/2)v^2} dv \tag{11.106}$$

where A is a similar constant to B and the two integrals have the same form. The integrals are closely related to Fresnel's integral defined as the following

$$\int_0^{u_1} e^{j(\pi/2)u^2} du = C(u_1) + jS(u_1) \tag{11.107}$$

$$\left.\begin{aligned} C(u_1) &= \int_0^{u_1} \cos\left(\frac{\pi}{2}u^2\right) du \\ S(u_1) &= \int_0^{u_1} \sin\left(\frac{\pi}{2}u^2\right) du \end{aligned}\right\} \tag{11.108}$$

These quantities are illustrated in the curve of Figure 11.36, known as Cornu's spiral, where u is the arc length from the origin.

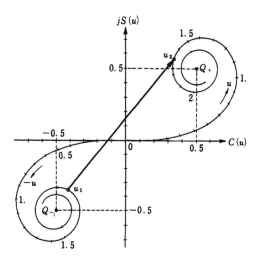

Figure 11.36 Cornu's spiral.

When $u_1 \to \pm\infty$,

$$\left.\begin{array}{l} C(\pm\infty) = \pm\tfrac{1}{2} \\ S(\pm\infty) = \pm\tfrac{1}{2} \end{array}\right\}$$

(11.109)

Thus, the values converge to the points $Q\pm$.
  Further

$$\int_{u_1}^{u_2} e^{j(\pi/2)u^2}\, du$$

expresses a vector connecting two points u_1 and u_2 on the curve. This means that amplitude and phase are indicated simultaneously.
  Then, assuming that S and P are situated in free space without any barrier, $u_1 = -\infty$, $v_1 = -\infty$, $u_2 = +\infty$, $v_2 = +\infty$, and Equation (11.106) becomes

$$U_0 = -jA\{2C(\infty) + 2jS(\infty)\}^2 = -jA(1+j)^2 = 2A$$

Therefore, the diffraction factor [DF], which is the ratio of the sound pressure when a screen exists to the sound pressure in free space, may be expressed as follows

$$\begin{aligned} [\mathrm{DF}] = \frac{U(p)}{U_0} &= \frac{-j}{2} \int_{u_1}^{u_2} e^{j(\pi/2)u^2}\, du \int_{u_1}^{u_2} e^{j(\pi/2)v^2}\, dv \\ &= \frac{-j}{2} [\{C(u_2) - C(u_1) + j(S(u_2) - S(u_1))\} \\ &\quad \times \{C(v_2) - C(v_1) + j(S(v_2) - S(v_1))\}] \end{aligned}$$

(11.110)

B. Semi-infinite thin screen

When a semi-infinite thin screen exists between S and P, the integration of Equation (11.106) is carried out over the remaining semi-infinite open surface. In this case, in Figure 11.35, $\xi_1 \to -\infty$, $\xi_2 \to +\infty$ and $\eta_2 \to +\infty$, so that, in Equation (11.110), $u_1 = -\infty$, $u_2 = +\infty$ and $v_2 = +\infty$, and hence

$$[\mathrm{DF}] = \frac{-j}{2}\{1+j\}\left\{(\tfrac{1}{2} - C(v_1)) + j(\tfrac{1}{2} - S(v_1))\right\}$$

(11.111)

The attenuation caused by the screen in this condition $[\mathrm{Att}]_{1/2}$, in decibel form, is thus

$$\begin{aligned} [\mathrm{Att}]_{1/2} &= -10\log_{10} |[\mathrm{DF}]|^2 \\ &= -10\log_{10} \tfrac{1}{2}\left\{(\tfrac{1}{2} - C(v_1))^2 + (\tfrac{1}{2} - S(v_1))^2\right\} \text{ (dB)} \end{aligned}$$

(11.112)

where the bracketed terms correspond to the square of the absolute value of the vector from v_1 to Q_+ on Cornu's spiral, as shown in Figure 11.36. The graph of $[\mathrm{Att}]_{1/2}$ vs v in Equation (11.112) is shown in Figure 11.37.

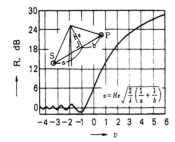

Figure 11.37 Kirchhoff's approximate theory of diffraction.

C. Interpretation of Fresnel's integral

It is necessary to understand the geometrical meaning of the integral variable v. For this purpose, use of the concept of Fresnel zones is appropriate.

Consider a spherical surface whose centre is at a sound source S with radius r_0, as shown in Figure 11.38. According to Huygens' principle, the sound wave reaching the point P must be a synthesis of secondary wavelets emitted from the spherical surface. Taking the reference point as M_0 on the line SP, the wavelets emitted from other points are reduced in amplitude due to the path difference δ, and their phases delayed by $k\delta(k = 2\pi/\lambda)$. When the secondary wavelets are expressed in the form of vectors, addition of these vectors leads to Cornu's spiral.

When a tangent is drawn at a point v on the spiral, making an angle ψ with the $C(v)$ axis, as shown in Figure 11.39, from Equation (11.108),

$$\tan \psi = \frac{dS(v)}{dC(v)} = \frac{dS(v)/dv}{dC(v)/dv} = \frac{\sin((\pi/2)v^2)}{\cos((\pi/2)v^2)} = \tan\left(\frac{\pi}{2}v^2\right)$$

$$\therefore \ \psi = \frac{\pi}{2}v^2$$

(11.113)

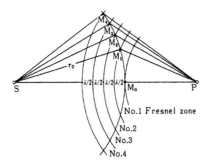

Figure 11.38 Fresnel zones.

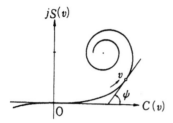

Figure 11.39 Detail of Cornu's spiral.

Taking the secondary wave vector from M_0 as the origin, put $\psi_{M0} = 0$. Then the phase lag ψ due to the path difference δ is

$$\psi = \frac{\pi}{2}v^2 = \frac{2\pi}{\lambda}\delta \quad \therefore \quad v = 2\sqrt{\frac{\delta}{\lambda}} \tag{11.114}$$

Hence it is found that v is related to the path difference δ and the wavelength λ.

In Figure 11.38, by plotting $M_1, M_2, \ldots, M_n$ on the spherical wave surface with the path differences $\delta_n = n \cdot \lambda/2$, $n = 1, 2, 3, \ldots$, then divided annular regions $M_0 - M_1$, $M_1 - M_2$, ... are created, which are called Fresnel's zones. So, in order to express the relationship between δ and λ the following general parameter is introduced,

$$N = \delta \cdot \frac{2}{\lambda} \tag{11.115}$$

called Fresnel's zone number (or Fresnel number). From Equations (11.114) and (11.115) we obtain $N = v^2/2$, and, changing the variable with the aid of this relation, Figure 11.37 can be rearranged to provide the dotted line values in Figure 5.15, with parameter N in a form which is convenient for practical calculations. In practice, in place of Equation (11.112), it is more convenient and reliable to use the values from the solid line in Figure 5.15, since it has been corrected by experiment.

The purpose of the design chart of Figure 5.15, is to make it easy to calculate the noise reduction by a screen without using a computer, as was the case in those days. Now, engineers everywhere use personal computers, so many acousticians have developed programs for personal computers that can calculate the values of Figure 5.15. The representative one is as follows

$$[\text{ATT}] = \begin{bmatrix} 10\log N + 13 & \text{for } N > 1 \\ 5 \pm 8|N|^{0.438} & \text{for } -0.34 < N < 1 \\ 0 & \text{for } N < -0.34 \end{bmatrix} \tag{11.116}$$

(Yamamoto & Takagi 1992)

Here, these values are valid for a semi-infinite thin screen in an infinite space without any obstacle. There should be no misunderstanding – if the ground is included, the above will not apply. This error was made in Final Report (Int. INCE 1999).

D. Screen of finite length

As shown in Figure 11.40, if the surface not obstructed by the screen is divided into region [A], which is semi-infinite, and regions [B] and [C], which are quarter-infinite, and integrating Equation (11.110) for each region, then the results can be combined. In the case of noise, their energies are summed, neglecting their phases, since they are incoherent. While, in region [A], Equation (11.112) is used, for a quarter-infinite zone, substituting $u_2 = +\infty$, $v_2 = +\infty$ into Equation (11.110) the attenuation denoted by $[\text{Att}]_{1/4}$ is obtained

$$
\begin{aligned}
[\text{Att}]_{1/4} = &-10\log_{10}\frac{1}{2}\left\{\left(\frac{1}{2}-C(u_1)\right)^2 + \left(\frac{1}{2}-S(u_1)\right)^2\right\} \\
&-10\log_{10}\frac{1}{2}\left\{\left(\frac{1}{2}-C(v_1)\right)^2 + \left(\frac{1}{2}-S(v_1)\right)^2\right\}
\end{aligned}
\tag{11.117}
$$

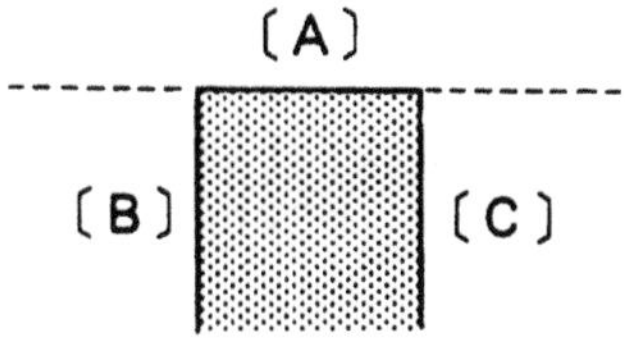

Figure 11.40 Screen of finite length.

Thus, the procedure has involved the summation of two terms of the same form as Equation (11.112). Figure 5.15 is applicable in either case. Therefore, an approximate calculation for diffraction due to a finite screen is possible by simply taking the values from the solid line in Figure 5.15 and summing them.

E. Thickness and edge section of barriers

So far the thickness of the screen has been neglected. However, in practice it will have a finite thickness and the section of the edge that produces diffraction may be simply described as a wedge of a certain apex angle see Figure 11.41. The effects of these characteristics on sound attenuation by diffraction can be expressed as correction terms, which should be added to the approximate calculation described above. They are derived by taking the difference between the approximate approach and a rigorous solution of the wave theory.

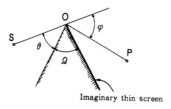

Imaginary thin screen

Figure 11.41 Diffraction by a wedge.

a. Effect of wedge angle

$$[Att]_\Omega = [Att]_0 + [EW]_\Omega \tag{11.118}$$

where $[EW]_\Omega$ is the effect of wedge angle Ω. The first term is obtained from Figure 5.15, while the second term can be obtained from Figure 11.42 (Maekawa & Osaki 1985).

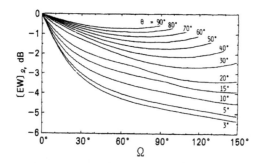

Figure 11.42 Effect of wedge angle Ω on attenuation by diffraction with source angle θ as parameter (Maekawa & Osaki 1985).

b. Effect of thickness of screen

In the same way as in the foregoing section, an imaginary thin screen is assumed on the surface on the receiver side of a thick barrier, as shown in Figure 11.43. The effect of thickness b on the noise attenuation $_n[ET]_b$ is obtained as a correction term, so that the noise attenuation by the thick barrier is

$$_n[Att]_b = [Att]_0 + _n[ET]_b \tag{11.119}$$

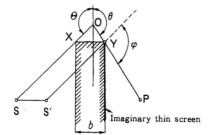

Figure 11.43 Diffraction by a thick barrier.

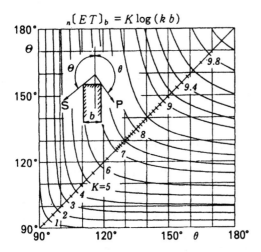

Figure 11.44 Chart for obtaining the effect of thickness b of a noise barrier.

where

$$_n[\text{ET}]_b = K\log_{10}(kb) \qquad (11.120)$$

K is obtained from Figure 11.44 (Fujiwara *et al.* 1977a).

F. *Effect of surface absorption*

All screens or barriers discussed above are assumed to have a rigid surface. But, barriers that have sound absorbent treatment are widely used. The effect of this is obtained from Figure 11.45, which is derived from exact theory with some approximation for simple conditions. The correction term is usually not large. This fact suggests that absorptive treatment is to be

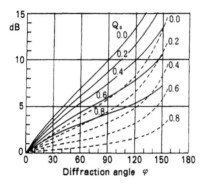

Figure 11.45 Effect of surface absorption on barrier attenuation calculated for a thin half-plane diffraction. Solid curves apply to a line source, dotted curves apply to a point source, Q_0 is the sound pressure reflection coefficient.

used, not for reducing the diffracted sound, but for suppressing the reflected sound (Fujiwara *et al.* 1977b).

11.15   Principle of statistical energy analysis (SEA)

Transmission of structure-borne sound in buildings is very difficult to solve analytically since most practical structures are complicated systems, each of which have resonant modes of vibration. So, a statistical approximate calculation of energy flow has been developed called statistical energy analysis (SEA).

Considering two systems I and II, coupled as shown in Figure 11.46, where P_{1in}, P_{2in} are the input powers, P_{1d}, P_{2d} are the dissipating powers and the power flow between the two systems is P_{12} from system I to system II, and P_{21} from system II to system I, respectively. Then

$$P_{1in} = P_{1d} - P_{12} + P_{21} \tag{11.121}$$

$$P_{2in} = P_{2d} - P_{21} + P_{12} \tag{11.122}$$

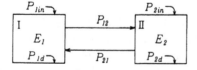

Figure 11.46 Energy flow between two coupled systems.

where the energy dissipated in each system P_{id} at angular frequency ω is

$$P_{id} = E_i \omega \eta_i \tag{11.123}$$

where E_i is the energy stored in one cycle of vibration and η_i is the dissipation loss factor in each system, respectively.

Assuming further, that the waves transmitting and exchanging energy through each system are not correlated with each other, we can separate the power flow as

$$P_{12} = E_1 \omega \eta_{12} \tag{11.124}$$
$$P_{21} = E_2 \omega \eta_{21} \tag{11.125}$$

where η_{12} is the coupling loss factor from system I to system II and η_{21} is the coupling loss factor from system II to system I.

Then the net power flow between the two systems is

$$\overline{P}_{12} = P_{12} - P_{21} = \omega \left(E_1 \eta_{12} - E_2 \eta_{21} \right) \tag{11.126}$$

Here we have assumed that each resonant mode within a narrow frequency band $\Delta\omega$ has the same energy, and also that the coupling of the individual resonant modes of both systems is approximately the same.

Now, defining the modal energy as

$$E_m = \frac{E(\Delta\omega)}{n(\omega)\,\Delta\omega} \left(\mathrm{W\,s\,Hz}^{-1} \right) \tag{11.127}$$

where $E(\Delta\omega)$ is the total energy in angular frequency band $\Delta\omega$ and $n(\omega)$ is the modal density, i.e. the number of modes in a band of unit width ($\Delta\omega = 1$) centred on ω.

From this definition, assuming an equal distribution of energy in the modes, and the same coupling loss factor,

$$\eta_{12} \cdot n_1(\omega) = \eta_{21} \cdot n_2(\omega) \tag{11.128}$$

therefore, combining Equations (11.126) and (11.128) gives

$$\overline{P}_{12} = \omega \eta_{12} n_1(\omega) \left[E_{1m} - E_{2m} \right] \Delta\omega (\mathrm{W}) \tag{11.129}$$

where E_{1m}, E_{2m} are the modal energies for systems I and II, respectively ($\mathrm{W\,s\,Hz}^{-1}$). This is the fundamental equation of the SEA method. It shows that the power flow is from the system with higher modal energy to that with lower modal energy. It does not depend on any dynamic variables, but is similar to the equation for heat flow due to a temperature difference.

Although it is a very simple equation, there are many problems to be investigated, such as the validity of the simplifying assumptions, the determination of loss factors, modal densities and modal energies and so on. (See Lit. B19, B22.)

11.16 Human subjective evaluation of an acoustic environment

Figure 11.47 shows a flowchart of the human subjective evaluation of an acoustic environment. In reality, things are more complicated, but, this simplified diagram is sufficient for the moment.

An acoustic signal $s(t)$ radiated from a sound source is affected by a room impulse response $r(t)$ and arrives at a listener. The composite acoustic signal is expressed as $s(t)*r(t)$. This composite acoustic signal is then affected by head-related impulse responses $h_{l,r}(t)$ as it arrives at the entrance of the right and the left ears as input signals to the auditory system. The input signals are expressed as $s(t)*r(t)*h_{l,r}(t)$.

In this the listener perceives auditory events, which include various groups of subjective attributes. The subjective attributes are divided into three groups. The first group concerns temporal attributes. The second group involves spatial attributes. The third relates to the quality attributes.

When the listener makes a subjective judgment of each subjective attribute, he/she is influenced by personal taste. In particular, the listener

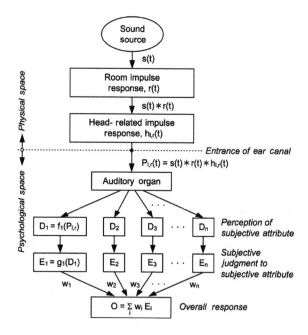

Figure 11.47 Flowchart of human subjective evaluation of an acoustic environment.

has an overall response to the acoustics of the space, which involves summing subjective judgments of each subjective attribute, weighted again by personal taste. In some cases, interactions occur.

Here, the overall response includes individual differences, because the subjective judgments to subjective attributes include individual differences. For instance, subjective preference, which is a kind of overall response, includes individual difference. On the other hand, the perception of individual subjective attributes does not include such individual differences.

For example, here are two cups of coffee. One contains no sugar and another contains two spoonfuls of sugar. The first question is 'Which is sweeter?' Surely, everyone answers: 'The latter is sweeter than the former.' If somebody answers with the opposite view, 'The former is sweeter than the latter,' it means only that his sensitivity to sweetness is very low.

This is the perception of a subjective attribute. The next question is 'Which do you prefer?' The answer depends on personal taste. It is no small wonder that subjects answer differently. This is overall response, a preference judgment.

Therefore, needless to say, it is impossible to control and evaluate overall responses of many unspecified listeners to the acoustics of a space (Morimoto 2001).

11.17 Outdoor planning of environmental acoustics

There are natural sounds in nature, even in the absence of humans. However, when people are present sounds are generated, and these sounds give information essential for human living. If a person loses his/her hearing, it will become clear to them that sound has a very important role for daily life. But with the expansion of social activities, not only necessary sounds but, increasingly, disturbing noise, threatens our social life because it is a noise nuisance. Therefore, in order to secure a quiet and comfortable environment, noise control engineering has been developed, targeting environmental standards.

However, suppressing the noise level below standard values does not guarantee a good acoustic environment, but is a passive device for protection from noise disturbance.

In determining what are good acoustics for the environment, many researchers in recent years have also considered natural sounds under the heading, soundscape. These investigations need not only natural sciences but also contributions from social sciences. Therefore, this is a multidisciplinary field beyond the scope of this book, which is limited to the acoustic engineering of physics and to hearing.

Appendices

Table A.1 Standing-wave ratio and related coefficients

20 log n (dB)	n	α_0 (%)	$|r_p|$ (%)
0.0	1.000	100.0	0.0
0.2	1.023	100.0	1.1
0.4	1.047	99.9	2.3
0.6	1.072	99.9	3.5
0.8	1.096	99.8	4.6
1.0	1.122	99.7	5.8
1.2	1.148	99.5	6.9
1.4	1.175	99.4	8.1
1.6	1.202	99.2	9.2
1.8	1.230	98.9	10.3
2.0	1.259	98.7	11.5
2.2	1.288	98.4	12.6
2.4	1.318	98.1	13.7
2.6	1.349	97.8	14.9
2.8	1.380	97.5	16.0
3.0	1.413	97.1	17.1
3.2	1.445	96.7	18.2
3.4	1.479	96.3	19.3
3.6	1.514	95.8	20.5
3.8	1.549	95.3	21.5
4.0	1.589	94.8	22.8
4.2	1.622	94.4	23.7
4.4	1.660	93.8	24.8
4.6	1.698	93.3	25.9
4.8	1.738	92.8	27.0
5.0	1.778	92.2	28.0
5.2	1.820	91.5	29.1
5.4	1.862	90.9	30.1
5.6	1.905	90.3	31.2
5.8	1.950	89.6	32.2
6.0	1.995	89.0	33.2
6.2	2.042	88.3	34.3
6.4	2.089	87.6	35.3
6.6	2.138	86.9	36.3
6.8	2.188	86.1	37.3

7.0	2.239	85.4	38.3
7.2	2.291	84.6	39.2
7.4	2.344	83.8	40.2
7.6	2.399	83.1	41.2
7.8	2.455	82.3	42.1
8.0	2.512	81.5	43.1
8.2	2.570	80.7	44.0
8.4	2.630	79.8	44.9
8.6	2.692	79.0	45.8
8.8	2.754	78.2	46.7
9.0	2.818	77.3	47.6
9.2	2.884	76.5	48.5
9.4	2.951	75.6	49.4
9.6	3.020	74.7	50.3
9.8	3.090	73.9	51.1
10.0	3.162	73.0	52.0
10.2	3.236	72.1	52.8
10.4	3.311	71.3	53.6
10.6	3.388	70.4	54.4
10.8	3.467	69.5	55.2
11.0	3.548	68.6	56.0
11.2	3.631	67.7	56.8
11.4	3.715	66.8	57.6
11.6	3.802	65.9	58.4
11.8	3.890	65.1	59.1
12.0	3.981	64.2	59.9
12.2	4.074	63.3	60.6
12.4	4.169	62.4	61.3
12.6	4.266	61.5	62.0
12.8	4.365	60.7	62.7
13.0	4.467	59.8	63.4
13.2	4.571	58.9	64.1
13.4	4.677	58.0	64.8
13.6	4.786	57.2	65.4
13.8	4.898	56.3	66.1
14.0	5.012	55.5	66.7
14.2	5.129	54.6	67.4
14.4	5.248	53.8	68.0
14.6	5.370	52.9	68.6
14.8	5.495	52.1	69.2
15.0	5.623	51.3	69.8
15.2	5.754	50.5	70.4
15.4	5.888	49.7	71.0
15.6	6.026	48.8	71.5
15.8	6.166	48.0	72.1
16.0	6.310	47.2	72.6
16.2	6.457	46.5	73.2
16.4	6.607	45.7	73.7
16.6	6.761	44.9	74.2
16.8	6.918	44.1	74.7
17.0	7.079	43.4	75.2
17.2	7.244	42.6	75.7
17.4	7.413	41.9	76.2

20 log n (dB)	n	α_0 (%)	$|r_p|$ (%)
17.6	7.586	41.2	76.7
17.8	7.762	40.4	77.2
18.0	7.943	39.7	77.6
18.2	8.128	39.0	78.1
18.4	8.318	38.3	78.5
18.6	8.511	37.6	79.0
18.8	8.710	37.0	79.4
19.0	8.913	36.3	79.8
19.2	9.120	35.6	80.2
19.4	9.333	35.0	80.6
19.6	9.550	34.3	81.0
19.8	9.772	33.7	81.4
20.0	10.000	33.1	81.8
20.5	10.59	31.5	82.7
21.0	11.22	30.0	83.6
21.5	11.89	28.6	84.5
22.0	12.59	27.3	85.3
22.5	3.34	25.9	86.1
23.0	14.13	24.7	86.8
23.5	14.96	23.5	87.5
24.0	15.85	22.3	88.1
24.5	16.79	21.2	88.8
25.0	17.78	20.2	89.4
25.5	18.84	19.2	89.9
26.0	19.95	18.2	90.5
26.5	21.13	17.3	91.0
27.0	22.39	16.4	91.5
27.5	23.71	15.5	91.9
28.0	25.12	14.7	92.3
28.5	26.61	14.0	92.8
29.0	28.18	13.2	93.2
29.5	29.85	12.5	93.5
30.0	31.62	11.9	93.9
31.0	35.48	10.7	94.5
32.0	39.81	9.6	95.1
33.0	44.67	8.6	95.6
34.0	50.12	7.7	96.1
35.0	56.23	6.9	96.5
36.0	63.10	6.2	96.9
37.0	70.79	5.5	97.2
38.0	79.43	4.9	97.5
39.0	89.13	4.4	97.8
40.0	100.0	3.9	98.0
42.0	125.9	3.1	98.4
44.0	158.5	2.5	98.8
46.0	199.5	1.9	99.0
48.0	251.2	1.6	99.2
50.0	316.2	1.3	99.4
55.0	562.3	0.7	99.6
60.0	1000.0	0.4	99.8

Table A.2 Sound absorption coefficients

2.A Sound absorption coefficients of common building materials

No.	Material	Frequency (Hz)					
		125	250	500	1k	2k	4k
1	Brick, bare concrete surface	0.01	0.02	0.02	0.02	0.03	0.04
2	Mortar smooth finish, plaster, marble, ceramic-tile finish	0.01	0.01	0.02	0.02	0.02	0.03
3	Cloth finish on concrete wall	0.03	0.03	0.03	0.04	0.06	0.08
4	Plastic tile on concrete floor	0.01	0.02	0.02	0.02	0.03	0.04
5	Wood floor (parquet or flooring on studs)	0.16	0.14	0.11	0.08	0.08	0.07
6	Needle-punch carpet	0.03	0.04	0.06	0.10	0.20	0.35
7	Pile carpet 10 mm thick	0.09	0.10	0.20	0.25	0.30	0.40
8	Glass wall, 10 mm thick	0.15	0.06	0.04	0.03	0.02	0.02
9	Window glass (in wooden frame)	0.35	0.25	0.18	0.12	0.07	0.04
10	Plexiglas for illuminating (910 × 910 × 2 mm)	0.40	0.25	0.20	0.20	0.22	0.25
11	Velvet curtain (with no drape)	0.05	0.07	0.13	0.22	0.32	0.35
12	Velvet curtain draped to half its area (100 mm air space)	0.10	0.25	0.55	0.65	0.70	0.70
13	Velvet curtain draped to half its area (500 mm air space)	0.15	0.25	0.50	0.75	0.80	0.85
14	Gravel, 150 mm thick	0.15	0.30	0.80	0.42	0.61	0.72
15	Sand (dried), 125 mm thick	0.24	0.34	0.45	0.62	0.76	0.95
16	Water surface	0.0l	0.01	0.01	0.02	0.02	0.03
17	Reflecting panels (of plywood with damping sheet) on stage	0.20	0.13	0.10	0.07	0.06	0.06
18	Equivalent absorption coefficient of virtual surface without the reflecting panels on stage	0.40	0.50	0.60	0.60	0.60	0.60
19	Proscenium opening	0.30	0.35	0.40	0.45	0.50	0.55
20	Lighting opening (of non-absorptive treated booth)	0.10	0.15	0.20	0.22	0.25	0.30
21	Opening for ventilation ducts, etc.	0.75	0.80	0.80	0.80	0.85	0.85
		Absorption Area (m^2)					
22	Plywood chair	0.02	0.02	0.02	0.04	0.04	0.03
23	Theatre chair covered with vinyl leather	0.04	0.13	0.22	0.17	0.16	0.11
24	Theatre chair upholstered with moquette	0.14	0.25	0.30	0.30	0.30	0.30
25	Person sitting on upholstered chair	0.25	0.35	0.40	0.40	0.40	0.40

2.B Sound absorption coefficients of porous materials

No.	Material	Thickness (mm)	Air layer (mm)	Frequency (Hz)					
				125	250	500	1k	2k	4k
1	Glass wool 16–24 kg m^{-3}	25	0	0.10	0.30	0.60	0.70	0.80	0.85
2		50	0	0.20	0.65	0.90	0.85	0.80	0.85
3		100	0	0.60	0.95	0.95	0.85	0.80	0.90
4		25	40	0.15	0.40	0.70	0.85	0.90	0.95
5		50	40	0.25	0.80	0.95	0.90	0.85	0.90
6		25	100	0.22	0.57	0.83	0.82	0.90	0.90
7		50	100	0.45	0.97	0.99	0.85	0.80	0.92
8		25	300	0.65	0.70	0.75	0.80	0.75	0.85
9		5	300	0.75	0.85	0.85	0.80	0.80	0.85
10	Glass wool 32–48 kg m^{-3}	25	0	0.12	0.30	0.65	0.80	0.85	0.85
11		50	0	0.20	0.65	0.95	0.90	0.80	0.85
12		25	40	0.12	0.45	0.85	0.90	0.85	0.90
13		50	40	0.28	0.90	0.95	0.87	0.85	0.94
14		25	100	0.25	0.70	0.90	0.85	0.85	0.90
15	Glass wool 32–40 kg m^{-3}	100	0	0.70	1.00	0.98	0.85	0.70	0.80
16		100	40	0.78	1.00	0.99	0.94	0.90	0.90
17	(with glass cloth cover)	100	100	0.80	1.00	0.99	0.93	0.84	0.84
18	Rock wool 40–140 kg m^{-3}	25	0	0.10	0.30	0.70	0.80	0.80	0.85
19		50	0	0.20	0.65	0.95	0.90	0.85	0.90
20		25	40	0.20	0.65	0.90	0.85	0.80	0.80
21		50	40	0.35	0.85	0.95	0.90	0.85	0.85
22		25	100	0.35	0.65	0.90	0.85	0.85	0.80
23		50	100	0.55	0.90	0.95	0.90	0.85	0.85
24		25	300	0.65	0.85	0.85	0.80	0.80	0.85
25		50	300	0.75	0.95	0.95	0.85	0.85	0.90
26	Soft urethane foam	20	0	0.07	0.20	0.40	0.55	0.70	0.70
27		20	40	0.10	0.25	0.60	0.90	0.80	0.85
28	Polystyrene foam	25	0	0.04	0.05	0.06	0.14	0.30	0.25
29	Sprayed rock wool	12	0	0.05	0.12	0.40	0.55	0.70	0.75
30		25	0	0.12	0.35	0.80	0.88	0.85	0.90
31	Sprayed pearlite plaster	5	0	0.04	0.10	0.17	0.17	0.19	0.20
32	Cemented fine excelsior board	25	0	0.03	0.14	0.30	0.55	0.65	0.60
33	Cemented chipped wood board	50	0	0.15	0.20	0.70	0.80	0.70	0.85

34	Rock wool board	12	300	0.35	0.30	0.40	0.55	0.65	0.70
35	(over gypsum board)	12	300	0.20	0.20	0.40	0.60	0.70	0.75
36	Ceramic board 1.36 g cm^{-3}	20	0	0.05	0.10	0.20	0.40	0.80	0.60
37	(with air space behind)	20	200	0.40	0.90	0.80	0.55	0.55	0.70

2.C Sound absorption coefficients of board-form materials

No.	Material	Thickness (mm)	Air layer (mm)	Frequency (Hz)					
				125	250	500	1k	2k	4k
1	Gypsum board	9–12	45	0.26	0.13	0.09	0.05	0.05	0.05
2	Flexible cement board	3–5	90	0.23	0.12	0.08	0.06	0.05	0.05
3	Calcium silicate board	6–8	180	0.18	0.10	0.08	0.06	0.06	0.05
4	Lauan plywood	3	45	0.45	0.16	0.10	0.08	0.07	0.08
5		6–9	45	0.15	0.28	0.12	0.07	0.07	0.08
6		6–9	90	0.25	0.17	0.09	0.07	0.07	0.08
7		12	45	0.25	0.14	0.07	0.04	0.07	0.08
8	Particle wood board	20	45	0.27	0.08	0.08	0.06	0.08	0.07
9	Corrugated polyester board		90	0.26	0.41	0.22	0.12	0.10	0.15

Table A.2 (Continued)

2.D Slit-resonator type sound absorbing structures

No.		Slit Width (mm)	Air Space (mm)	Frequency (Hz) 125	250	500	1k	2k	4k
1	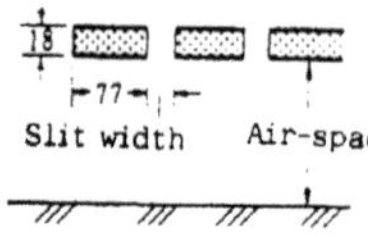18 77 Slit width Air-space	4.5	80	0.08	0.35	0.22	0.15	0.15	0.15
2		4.5	180	0.22	0.33	0.17	0.24	0.14	0.12
3		20.0	80	0.07	0.15	0.33	0.20	0.18	0.17
4		20.0	180	0.12	0.37	0.28	0.25	0.17	0.15
5		4.5	80	0.18	0.77	0.40	0.24	0.20	0.26
6	Cotton Cloth Lining {	20.0	80	0.12	0.50	0.68	0.40	0.26	0.22
7	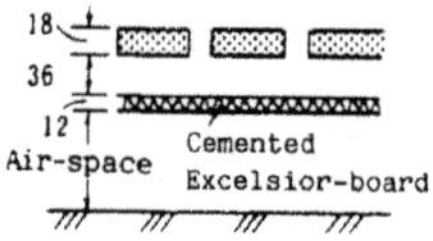18 36 12 Air-space Cemented Excelsior-board	4.5	50	0.15	0.37	0.40	0.27	0.15	0.12
8		20.0	50	0.10	0.25	0.44	0.42	0.35	0.30
9		20.0	300	0.45	0.50	0.37	0.40	0.40	0.37
10	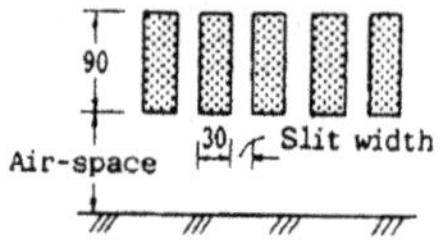90 30 Slit width Air-space	10.0	50	0.08	0.20	0.30	0.24	0.35	0.38
11		10.0	300	0.37	0.40	0.32	0.40	0.40	0.44
12	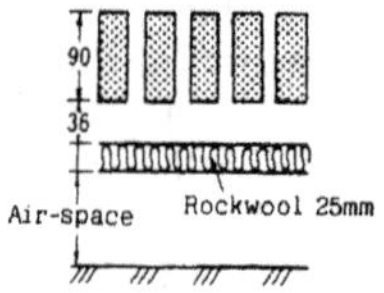90 36 Air-space Rockwool 25mm	10.0	50	0.25	0.87	0.55	0.40	0.50	0.44
13		10.0	300	0.68	0.70	0.67	0.64	0.52	0.50
14		4.5	50	0.65	0.65	0.50	0.40	0.40	0.40
15		30.0	50	0.20	0.60	0.75	0.65	0.70	0.60

Table A.2 (Continued)

2.E Sound absorption coefficients of perforated structures

Diameter pitch (perforation) No. (rate)	Thickness of board (mm)	Materials under the board (thickness) (mm)	Air layer (mm)	Frequency (Hz)					
				125	250	500	1k	2k	4k
1 4_φ–15 (6%)	5	non	45	0.02	0.09	0.25	0.31	0.15	0.10
2		Glass wool (25)	45	0.15	0.35	0.82	0.52	0.23	0.22
3		non	180	0.12	0.45	0.30	0.25	0.23	0.16
4		non	500	0.45	0.31	0.31	0.30	0.30	0.28
5		Glass wool (25)	500	0.87	0.61	0.70	0.65	0.46	0.33
6 6_φ–22 (6%)	9	non	45	0.03	0.09	0.46	0.31	0.18	0.15
7		Rock wool (25)	45	0.09	0.50	0.94	0.44	0.22	0.21
8		non	300	0.35	0.37	0.25	0.22	0.23	0.22
9		Rock wool (25)	300	0.68	0.82	0.58	0.53	0.33	0.23
10 6_φ–15 (13%)	5	non	45	0.02	0.08	0.16	0.31	0.20	0.18
11		Glass wool (50)	45	0.13	0.32	0.78	0.69	0.40	0.31
12		non	500	0.35	0.29	0.30	0.35	0.36	0.39
13		Glass wool (25)	500	0.87	0.68	0.76	0.82	0.71	0.50
14 8_φ–16 (20%)	4	non	300	0.22	0.30	0.29	0.23	0.19	0.28
15		Glass wool (25)	300	0.61	0.73	0.64	0.61	0.62	0.58
16		Rock wool (25)	300	0.85	0.94	0.83	0.75	0.66	0.60
17 9_φ–15 (28%)	5	non	45	0.01	0.05	0.11	0.21	0.16	0.13
18		Glass wool (50)	45	0.15	0.30	0.68	0.78	0.59	0.58
19	6	Rock wool (25)	45	0.08	0.25	0.71	0.91	0.78	0.72
20	5	non	500	0.30	0.25	0.27	0.36	0.39	0.42
21		Glass wool (25)	500	0.83	0.72	0.80	0.90	0.87	0.70

Table A.3 Average transmission loss of various constructions

Table A.3 (Continued)

3.A Transmission loss of single boards (dB)

No.	Material	Thickness (mm)	Surface density $(kg\,m^{-2})$	Frequency (Hz) 125	250	500	1k	2k	4k	Average 125–4k
1	Lauan plywood	6	3.0	11	13	16	21	25	23	17
2		12	8.0	18	20	24	24	25	30	22
3		40	24	24	25	27	30	38	43	29
4	Particle wood board (Homogeneous)	20	13	24	27	26	27	24	33	26
5	(Novopan)	35	17	21	23	27	28	24	29	25
6	Pearlite board	12		17	18	24	30	33	30	24
7	Flexible cement board	4	7.1	18	22	23	28	33	36	25
8		6	11	19	25	25	31	34	28	27
9	Gypsum board	9	8.7	15	20	25	28	34	25	25
10	Sheet metal, aluminium	1.2		8	11	14	21	27	30	18
11	Sheet metal, steel	0.7	5.6	9	14	20	26	30	37	20
12	Sheet metal, lead	1.0	11.3	25	25	29	33	38	43	32
13	Glass pane	3	7.5	15	18	22	28	32	24	23
14		6	15	17	23	28	29	25	36	26
15		10	25	21	27	31	29	33	42	31
16	Double glazing: 6–18 mm airspace between double panes of 5 mm	16		16	17	27	32	28	30	26
		~28		~18	~21	~31	~34	~30	~32	
17	Glass block 145 × 145	95	97	30	32	38	46	53	39	40
18	Ordinary concrete, surface unfinished	120		32	40	46	53	59	64	49
19		150		35	40	49	55	60	65	51
20	Foamed concrete (ALC)	150		33	33	30	42	50	55	41
21	ditto resin-plaster, 3 mm both sides	150		31	33	40	46	52	56	43
22	Concrete block, bare surface	100	160	19	24	28	32	36	40	28
23	ditto oil-paint both sides	100	160	32	36	40	48	54	51	41
24	ditto plaster,15 mm both sides	100	160	33	37	42	49	56	57	43
25	ditto mortar finish both sides	150	180	31	35	45	52	56	56	44

Table A.3 (Continued)

3.B Transmission loss of double leaf walls (dB)

No.	Material and Structure	Surface density $(kg\,m^{-2})$	Frequency (Hz)						Average
			125	250	500	1k	2k	4k	125–4k
Surface material (thickness– airlayer– thickness) + absorbing materials inside (mm)									
1	Plywood $5-75-5$		13	17	25	31	40	40	28
2	Ditto + glass wool 25		20	17	31	36	46	47	33
3	Novopan $8-45-8$		5	11	22	31	37	33	20
4	$8-100-8$		14	21	31	40	50	36	31
5	Gypsum board $9-42-9$		15	22	27	33	30	32	28
6	ditto + glass wool 25		16	28	32	37	33	34	32
7	Gypsum board $12-65-12$	18.6	15	21	30	40	40	42	33
8	Ditto + rock wool 50	20.6	15	32	48	54	50	45	43
9	Gypsum board $12 \times 2-65-12 \times 2$	39.2	20	33	40	50	50	50	43
10	Heavy gypsum board + GW50	49.0	30	40	50	57	55	52	50
11	Gypsum board $15 \times 2-65-15 \times 2$	43.6	25	35	43	53	50	55	45
12	GW-Gypsum board $15 \times 3-65-15 \times 3$	75.0	28	40	49	57	57	56	50
13	Calcium silicate board $6+8-75-8+6$	42.6	27	35	43	48	53	59	45
14	ditto $8 \times 3-65-8 \times 3$	43.0	29	41	47	51	54	54	47

Table A.3 (Continued)

3.C Transmission loss of sandwich panels (dB)

No.	Material and construction	Frequency (Hz)						Average
		125	250	500	1k	2k	4k	125–4k
	Surface material (thickness – that of core – that of surface) Core Material							
1	Plywood (6–50–6) polyurethane foam	14	18	20	16	32	32	22
2	Plywood (3–60–3) polystyrene foam	19	21	24	22	30	38	24
3	Gypsum board (7–50–7) polyurethane foam	16	20	21	22	40	45	27
4	Fibre cement slate (6–50–6) polyurethane foam	20	21	19	30	37	44	28
5	Flexible cement board (3–20–3) polyurethane foam	21	18	23	24	22	45	25
6	ditto (3–25–3) cemented excelsior board	26	25	27	30	35	40	30
7	ditto (3–30–3) foamed concrete board	30	32	33	35	43	47	36
8	ditto (3–40–3) pearlite grain	26	27	29	31	36	38	31
9	ditto (5–40–5) glass wool	21	26	35	38	45	37	33
10	Plywood (4–25–4) paper honeycomb	11	13	16	18	22	30	18
11	Gypsum board (9–20–9) ditto	15	22	27	30	32	31	26
12	Aluminium plate (1–25–1) ditto	16	15	17	17	21	23	18

Table A.3 (Continued)

3.D Transmission loss of window and openings (dB)

No.	*Material and construction*	*Frequency (Hz)* 125	250	500	1k	2k	4k	*Average* 125–4k
1	Popular sliding aluminium sash (glass 5 mm)	18	20	23	21	22	25	22
2	Airtight aluminium sash (glass 5 mm)	20	23	27	29	29	30	26
3	Sliding aluminium sash double (air layer 100 mm)	18	22	26	25	21	32	24
4	ditto (air layer 200 mm)	25	27	32	26	24	32	27
5	Airtight aluminium sash double (air layer 100 mm)	27	31	31	34	34	36	32
6	ditto (air layer 200 mm)	27	34	40	45	50	50	41
7	2 mm steel flash door with air layer 45 mm	25	30	34	37	36	35	32
8	Aluminium insulation door for studio use	30	42	45	47	57	55	45

Table A.4 Physical dimensions and units

	MKS → cgs	*cgs → MKS*	*Dimension*
Density	$1\,\mathrm{kg\,m^{-3}} = 10^{-3}\,\mathrm{g\,cm^{-3}}$	$1\,\mathrm{g\,cm^{-3}} = 10^{3}\,\mathrm{kg\,m^{-3}}$	ML^{-3}
Surface density	$1\,\mathrm{kg\,m^{-2}} = 10^{-1}\,\mathrm{g\,cm^{-2}}$	$1\,\mathrm{g\,cm^{-2}} = 10\,\mathrm{kg\,m^{-2}}$	ML^{-2}
Speed	$1\,\mathrm{m^{-}s} = 10^{2}\,\mathrm{cm^{-}s}$	$1\,\mathrm{cm^{-}s} = 10^{-2}\,\mathrm{m^{-}s}$	LT^{-1}
Acceleration	$1\,\mathrm{m\,s^{-2}} = 10^{2}\,\mathrm{cm\,s^{-2}}$	$1\,\mathrm{cm\,s^{-2}} = 10^{-2}\,\mathrm{m\,s^{-2}}$	LT^{-2}
Force	$1\,\mathrm{N} = 10^{5}\,\mathrm{dyne}$ $(\equiv 1\,\mathrm{kg\,m\,s^{-2}})$	$1\,\mathrm{dyne} = 10^{-5}\,\mathrm{N}$ $(\equiv 1\,\mathrm{g\,cm\,s^{-2}})$	MLT^{-2}
Pressure (sound pressure)	$1\,\mathrm{N\,m^{-2}} = 10\,\mathrm{dyne\,cm^{-2}}$ $(\equiv 10\,\mu\mathrm{bar}) = 1\,\mathrm{Pa}$	$1\,\mathrm{dyne\,cm^{-2}} = 10^{-1}\,\mathrm{N\,m^{-2}}$ $(\equiv 1\,\mu\mathrm{bar})$	$ML^{-1}T^{-2}$
Work load (sound energy)	$1\,\mathrm{J} = 10^{7}\,\mathrm{erg}(\equiv 1\mathrm{N\,m})$	$1\,\mathrm{erg} = 10^{-7}\,\mathrm{J}(\equiv 1\,\mathrm{dyne\,cm})$	$ML^{2}T^{-2}$
Power (acoustic power)	$1\,\mathrm{W} = 10^{7}\,\mathrm{erg\,s^{-1}}(\equiv 1\,\mathrm{J\,s^{-1}})$	$1\,\mathrm{erg\,s^{-1}} = 10^{-7}\,\mathrm{w}$	$ML^{2}T^{-3}$
Acoustic impedance-density or Specific-acoustic impedance (flow resistance)	1 MKS rayl = 10 rayl $(\equiv 1\,\mathrm{N\,s\,m^{-3}} \equiv 1\,\mathrm{kg\,m^{-2}\,s^{-1}})$	1 rayl = 10^{-1} MKS rayl $(\equiv 1\,\mathrm{dynes\,cm^{-3}} \equiv 1\,\mathrm{g\,cm^{-2}\,s^{-1}})$	$ML^{-2}T^{-1}$
Acoustic resistance (ac. ohm)	1 MKS ac.Ω = 10^{3} cgs ac.Ω (≡ 1 MRS rayl.m^{2} ≡ $\mathrm{kg\,s^{-1}}$)	1 cgs ac.Ω = 10^{-3}MKS ac.Ω (≡ 1rayl $\mathrm{cm^{2}} \equiv 1\,\mathrm{g\,s^{-1}}$)	MT^{-1}

Table A.5 A, C, and G weighting in 1/3 octave bands

Frequency (Hz)	A-weighting (dB)	C-weighting (dB)	G-weighting (dB)
4			−16.0
5			−12.0
6.3			−8.0
8			−4.0
10	−70.4	−14.3	0.0
12.5	−63.4	−11.2	4.0
16	−56.7	−8.5	8.0
20	−50.5	−6.2	9.0
25	−44.7	−4.4	4.0
31.5	−39.4	−3.0	−4.0
40	−34.6	−2.0	−12.0
50	−30.2	−1.3	−20.0
63	−26.2	−0.8	−28.0
80	−22.5	−0.5	−36.0
100	−19.1	−0.3	−44.0
125	−16.1	−0.2	
160	−13.4	−0.1	
200	−10.9	0.0	
250	−8.6	0.0	
315	−6.6	0.0	
400	−4.8	0.0	
500	−3.2	0.0	
630	−1.9	0.0	
800	−0.8	0.0	
1000	0.0	0.0	
1250	0.6	0.0	
1600	1.0	−0.1	
2000	1.2	−0.2	
2500	1.3	−0.3	
3150	1.2	−0.5	
4000	1.0	−0.8	
5000	0.5	−1.3	
6300	−0.1	−2.0	
8000	−1.1	−3.0	
10 000	−2.5	−4.4	
12 500	−4.3	−6.2	
16 000	−6.6	−8.5	
20 000	−9.3	−11.2	

Bibliography

Reference papers

Ando Y. (1965) *Proc. Ann. Meeting Acoust. Soc. Jpn.*, Oct. 1965: 89–90 (in Japanese).

Ando Y. (1966) *Proc. Ann. Meeting Acoust. Soc. Jpn.*, May 1966: 219–220 (in Japanese).

Ando Y. (1968) *IEEE: Electro. Commun. Jpn.* 51A-8: 303–310.

Ando Y. & Hattori H. (1970) *J. Acoust. Soc. Am.* 47(4): 1128–1130.

Ando Y. & Hattori H. (1977) *Br. J. Obstet. Gynaecol.* 84(2): 115–118.

Barron M. (1971) *J. Sound Vib.* 15: 475–494.

Barron M. & Chiney C. B. (1979) *Appl. Acoust.* 12(5): 361–375.

Barron M. & Marshall A. H. (1981) *J. Sound Vib.* 77: 211–232.

Baulac M., Defrance J., & Jean P. (2008) *Appl. Acoust.* 69: 332–342.

Beranek L. L. *et al.* (1946) *J. Acoust. Soc. Am.* 18(1): 140–150.

Beranek L. L. (1947) *J. Acoust. Soc. Am.* 19: 556–568.

Beranek L. L. (1957) *Noise Control* 3: 19–27.

Berkhout A. J. (1988) *J. Aud. Eng. Soc.* 36(12): 977–995.

Bolt R. H. & Doak P. E. (1950) *J. Acoust. Soc. Am.* 22: 507–509.

Brebeck D., Bücklein R., Krauth E., Spandöck F. *et al.* (1967) *ACUSTICA* 18: 213–226.

Breeuwer R. & Tukker J. C. (1976) *Appl. Acoust.* 9(2): 77–101.

Cadoux R. E. (1992) *Euro. Noise '92, Proc. I.OA.* 14(pt. 4): 41–47.

Chrisler V. L. (1934) *J. Acoust. Soc. Am.* 5: 220.

Chu W. T. (1978) *J. Acoust. Soc. Am.* 63: 1444–1450.

Cook R. K., Waterhouse R. V., Berendt R. D., Edelman S., & Thompson M. C. (1955) *J. Acoust. Soc. Am.* 27: 1072.

Corcione M., Cianfrini C. *et al.* (2007) *Appl. Acoust.* 68/11–12: 1357–1372.

Damaske P. (1967/68) *ACUSTICA* 19: 199–213.

Delany M. E. & Bazley E. N. (1970) *Appl. Acoust.* 3(2): 105–116.

Els H. & Blauert J. (1986) *Proc. Symp. Vancouver 12th ICA*: 65–70.

Eyring C. F. (1931) *J. Acoust. Soc. Am.* 3: 181.

Fujiwara K., Ando Y., & Maekawa Z. (1977a) *Appl. Acoust.* 10(2): 147–159.

Fujiwara K., Ando Y., & Maekawa Z. (1977b) *Appl. Acoust.* 10(3): 167–179.

Fujiwara K. & Miyajami T. (1992) *Appl. Acoust.* 35(2): 149–152.

Fukushima A. *et al.* (1990) *Proc. Ann. Meeting INCE/Jpn*, Oct. 1990: 37–40 (in Japanese).

Furukawa H., Fujiwara K., Ando Y., & Maekawa Z. (1990) *Appl. Acoust.* 29(4): 255–271.

Gomperts M. C. (1964) *ACUSTICA* 14: 1.

Gomperts M. C. & Kihlman T. (1967) *ACUSTICA* 18: 144.

Gutowski T. G. & Dym C. L. (1976) *J. Sound Vib.* 49(2): 179–193.

Hayakusa S. (1990) *Archi. Envi. Acoust.* by Maekawa Z. (Kyouritsu-Pub.) 153. (in Japanese).

Houtgast T. & Steeneken H. (1973) *ACUSTICA* 28: 66–73.

Hunt F. V. (1939) *J. Acoust. Soc. Am.* 10: 216–227.

Ingard U. & Bolt R. H. (1951) *J. Acoust. Soc. Am.* 23: 533–540.

Int. INCE (1999) NOISE/NEWS 7(3): 137–161.

Isei T., Embleton T. F. W., & Piercy J. E. (1980) *J. Acoust. Soc. Am.* 67(1): 46–58.

Ishii K. & Tachibana H. (1974) *Proc. 8th ICA (London)* 2: 610.

Kiyama M., Sakagami K., Tanigawa M., & Morimoto M. (1998) *Appl. Acoust.* 54(3): 239–254.

Konishi K., Aoki M. & Maekawa Z. (1979) *Proc. Inter-noise 1979 (Warzaw)*: 761–764.

Kosten C. W. (1960) *ACUSTICA* 10: 400.

Krokstad A., Strøm S., & Sørsdal S. (1983) *Appl. Acoust.* 16(4): 291–312.

Larrson C., Hallberg B., & Israelsson S. (1988) *Appl. Acoust.* 25(1): 17–31.

Leventhall H. G. (1987) *Proc. 4th Int. Meeting Low Freq. Noise Vib.* 2-5-1.

London A. (1950) *J. Acoust. Soc. Am.* 22: 263.

Maa D-Y. (1975) *Sci. Sin.* 17: 55–71.

Maa D-Y. (1998) *J. Acoust. Soc. Am.* 104(5): 2861–2866.

Maekawa Z. (1957) *Proc. Ann. Meeting Acoust. Soc. Jpn.*, May 1957: 49–50 (in Japanese).

Maekawa Z. (1959) *Proc. Ann. Meeting Acoust. Soc. Jpn.*, Nov. 1959: 71–72 (in Japanese).

Maekawa Z. (1962) *Proc. Ann. Meeting Acoust. Soc. Jpn.*, May 1962: 185–186 (in Japanese).

Maekawa Z. & Ando Y. (1964) *Proc. Ann. Meeting Archi. Inst. Jpn. 103*, Oct. 1964: 254 (in Japanese).

Maekawa Z. *et al.* (1965) *Proc. Ann. Meeting Acoust. Soc. Jpn.*, Oct. 1965: 191–192 (in Japanese).

Maekawa Z. Sakurai Y. and Ando Y. (1965) *Proc. Ann. Meeting Archi. Inst. Jpn.*, Sep. 1965, Extra: 385 (in Japanese).

Maekawa Z. (1968) *Appl. Acoust.* 1(3): 157–173.

Maekawa Z. (1968) *Architectural Acoustics* (Kyouritu-Shuppan) (in Japanese).

Maekawa Z. & Sakurai Y. (1968) *Proc. 6th ICA (Tokyo)* E-1-8.

Maekawa Z., Fujiwara K., Nagano N., & Morimoto M. (1971) *Symp. Noise Prevention 7th ICA. (Miskolc, Hungary)* 4.8: 1–7.

Maekawa Z. (1978) *Architectural Acoustics* (Kyouritu-Shuppan) (in Japanese).

Maekawa Z. & Osaki S. (1983) *Proc. 11th ICA (Paris)* 72: 1–4.

Maekawa Z. & Osaki S. (1985) *Appl. Acoust.* 18(5): 355–368.

Maekawa Z. (1990) *Architectural and Environmental Acoustics* (Kyouritu-Shuppan) (in Japanese).

Mechel F. P. (1988) *J. Acoust. Soc. Am.* 83(3): 1002–1013.

Miki Y. (1990) *J. Acoust. Soc. Jpn.* (E) 11(1): 19–24.

Miwa T. & Yonekawa Y. (1974) *Appl. Acoust.* 7(2): 83–101.

Monazzam M. R. & Lam Y. W. (2008) *Appl. Acoust.* 69: 93–104.

Morimoto M. (2001) *Proceedings of 2nd International Workshop on Spatial Media (Aizu-Wakamatsu, Japan)*, 1–15.

Nakashima T. & Isei S. (2004) *Proc. Inter-Noise (Prague).* 726.

Okubo T. & Fujiwara K. (1999) *J. Acoust. Soc. Am.* 3326–3335.

Okubo T., Matsumoto T., Yamamoto K., Funahashi O. & Nakasaki K. (2010) *J. Acoust. Sci. & Tech.* 31(1): 56–67.

Piercy J. E., Embleton T. F. W. & Sutherland L. C. (1977) *J. Acoust. Soc. Am.* 61(6): 1403–1418.

Pinnington R. J. & Nathanail C. B. (1993) *Appl. Acoust.* 40(1): 21–46.

Rindel J. H. (1991) *Appl. Acoust.* 34: 7–17.

Rindel J. H. (1993) *Appl. Acoust.* 38: 223–234.

Sakagami K., Kiyama M., Morimoto M., & Takahashi D. (1996) *Appl. Acoust.* 49(3): 237–247.

Sakamoto S., Nagatomo H., Ushijima A., & Tachibana H. (2008) *Acoust. Sci. Technol.* 29(4): 256–265.

Sakurai Y. (1987) *J. Acoust. Soc. Jpn.* (E) 8(4): 127–138.

Schroeder M. R. (1965) *J. Acoust. Soc. Am.* 38: 409–412.

Schroeder M. R. (1979) *J. Acoust. Soc. Am.* 65(4): 958–963.

Sekiguchi K., Kimura S., & Sugiyama T. (1985) *J. Acoust. Soc. Jpn.* (E) 6(2): 103–115.

Shioda M. (1986) *J. Low Freq. Noise Vib.* 5(2): 51–59.

Spandöck F. (1934) *Ann. Phys.* 20: 345.

Steinke G. (1983) *J. Audio Eng. Soc.* 31(7–8): 500–511.

Stevens S. S. (1972) *J. Acoust. Soc. Am.* 51(2): 575–593.

Strube H. W. (1981) *J. Acoust. Soc. Am.* 70(2): 633–635.

Tachibana H., Yamasaki Y., Morimoto M., Hirasawa Y., Maekawa Z. & Poesselt C. (1989) *J. Acoust. Soc. Jpn.* (E) 10(2): 73–85.

Tachibana H., Yamasaki Y. *et al.* (1989) *J. Acoust. Soc. Jpn.* (E) 10(2): 73–85.

Takagi K., Yamamoto T. & Maekawa Z. (1985) *Proc. Inter-noise 1985 (Munich)*: 473–476.

Terai T. & Kawai Y. (1990) *J. Acoust. Soc. Jpn.* (E) 11(1): 1–10.

Vercammen M. L. S. (1989) *J. Low Freq. Noise Vib.* 2: 105–109.

Vercammen M. L. S. (1992) *J. Low Freq. Noise Vib.* 11: 7–13.

Watters B. G. (1959) *J. Acoust. Soc. Am.* 31(7): 898–911.

West M., Sack R. A., & Walken F. (1991) *Appl. Acoust.* 33(3): 199–228.

Wilson G. P. & Soroka W. (1965) *J. Acoust. Soc. Am.* 37: 286.

Wu M. Q. (1997) *Noise Con. Eng. J.* 45(2): 69–77.

Yairi M., Sakagami K., Morimoto M., Minemura A., & Andow K. (2003) *Proceedings of 10th ICSV (Stockholm).*

Yamamoto K. & Takagi K. (1992) *Appl. Acoust.* 37(1): 75–82.

Yamasaki Y. & Itow T. (1989) *J. Acoust. Soc. Jpn.* (E) 10(2): 101–110.

Yamasaki Y. (2010), *Archi. Envi. Acoust.* by Z. Maekawa *et al.* (Kyouritsu-Pub.) (in Japanese).

Zwicker E. (1960) *ACUSTICA* 10: 304–308.

Zwicker E., Fastl H., Widmann U., Kurakata K., Kuwano S., & Namba S. (1991) *J. Acoust. Soc. Jpn.* (E) 12(1): 39–42.

Literature A: for acoustic design

1. V. O. Knudsen & C. M. Harris, *Acoustical Designing in Architecture* (1950, John Wiley; 1978, Acoust. Soc. Am.).
2. F. Ingerslev, *Acoustics in Modern Building Practice* (1952, Architectural press).
3. P. H. Parkin & H. R. Humphreys, *Acoustics Noise and Architecture* (1958, Faber & Faber).
4. L. L. Beranek, *Music Acoustics and Architecture* (1962, John Wiley).
5. A. Lawrence, *Architectural Acoustics* (1970, Elsevier).
6. L. L. Doelle, *Environmental Acoustics* (1972, McGraw-Hill).
7. ASHRAE Hand Book, *Systems Volume 1980* (The American Society of Heating, Refrigerating and Air-conditioning Engineers Inc.).
8. D. Collison, *Stage Sound* (1982, Cassell pub.).
9. R. H. Talaske, E. A. Wetherill, & W. J. Cavanaugh, eds., *Halls for Music Performance: Two Decades of Experience, 1962–1982* (1982, Acoust. Soc. Am.).
10. D. Lubman & E. A. Wetherill, eds., *Acoustics of Worship Spaces* (1985, Acoust. Soc. Am.).
11. M. Forsyth, *Buildings for Music* (1985, MIT Press).
12. R. H. Talaske & R. E. Boner, eds., *Theatres for Drama Performance* (1986, Acoust. Soc. Am.).
13. P. Lord & D. Templeton, *The Architecture of Sound, Designing Places of Assembly* (1986, Architectural Press).
14. A. Lawrence, *Acoustics and the Built Environment* (1989, Elsevier Science Pub.).
15. J. Eargle, *Handbook of Sound System Design* (1989, ELAR Pub.).
16. M. Barron, *Auditorium Acoustics and Architectural Design* (1993, E & FN Spon).
17. L. L. Makrinenko & J. S. Bradley, eds., *Acoustics of Auditoriums in Public Buildings* (1986, 1994, Acoust. Soc. Am.).
18. C. M. Harris, ed., *Noise Control In Buildings* (1994 McGraw-Hill, 1997, Inst. Noise Con. Eng.).
19. P. Lord & D. Templeton, *Detailing for Acoustics* (3rd ed. 1996, E & FN Spon).
20. L. L. Beranek, *Concert and Opera Halls, How They Sound* (1996, Acoust. Soc. Am.).
21. C. M. Harris, ed., *Handbook of Acoustical Measurements and Noise Control* (1998, Acoust. Soc. Am.).
22. W. Ahnert & F. Steffen, *Sound Reinforcement Engineering* (1994, 1999, E & FN Spon).
23. B. Kotzen & C. English, *Environmental Noise Barriers* (1999, E & FN Spon).
24. D. A. Bies & C. H. Hansen, *Engineering Noise Control, Theory and Practice* (3rd ed., 2003, E & FN Spon).
25. I. B. Hoffman, C. A. Storch, & T. J. Foulkes, eds., *Halls for Music Performance: Another Two Decades of Experience, 1982–2002* (2003, Acoust. Soc. Am.).
26. L. L. Beranek, *Concert Halls and Opera Houses. Music, Acoustics, and Architecture* (2nd ed., 2004, Springer).
27. M. Long, *Architectural Acoustics* (2006, Academic Press).

Literature B: for acoustic research

1. L. Rayleigh, *Theory of Sound I, II, 1877* (1945, Dover).
2. W. C. Sabine, *Collected Papers on Acoustics, 1923* (1964, Dover, 1993, Acoust. Soc. Am.).
3. H. Lamb, *The Dynamical Theory of Sound, 1925* (1960, Dover).
4. V. O. Knudsen, *Architectural Acoustics* (1932, John Wiley & Son).
5. S. S. Stevens & H. Davis, *Hearing its Psychology and Physiology 1938* (1983, Acoust. Soc. Am.).
6. P. M. Morse & R. H. Bolt, *Sound Waves in Rooms* (*Rev. Mod. Phys.* 16(2), April, 1944).
7. P. M. Morse, *Vibration and Sound* (2nd ed., 1948 McGraw-Hill, 1981, Acoust. Soc. Am.).
8. C. Zwicker & C. W. Kosten, *Sound Absorbing Materials* (1949, Elsevier).
9. L. L. Beranek, *Acoustic Measurements* (1949, John Wiley). Revised (1988, Am. Inst. Phys.).
10. P. V. Brüel, *Sound Insulation and Room Acoustics* (1951, Chapman Hall).
11. H. Fletcher, *Speech and Hearing in Communication* (1953, Van Nostrand, 1995, Acoust. Soc. Am.).
12. E. G. Richardson, ed., *Technical Aspect of Sound, Vol. I* (1953, Elsevier).
13. L. L. Beranek, *Acoustics* (1954, McGraw-Hill, 1987, Acoust. Soc. Am.).
14. C. M. Harris, ed., *Handbook of Noise Control* (1957, 2nd ed., 1979, McGraw-Hill).
15. H. F. Olson, *Acoustical Engineering* (1957, Van Nostrand).
16. L. L. Beranek, ed., *Noise Reduction* (1960, McGraw-Hill, 1991, Peninsula Pub.).
17. E. G. Richardson & E. Meyer, ed., *Technical Aspect of Sound, Vol. III* (1962, Elsevier).
18. G. Kurtze, *Physik und Technik der Lärmbekämpfung* (1964, G. Braun).
19a. L. Cremer & M. Heckl, *Körperschall* (1967, Springer-Verlag).
19b. L. Cremer & M. Heck, tr. by E. E. Ungar, *Structure-Borne Sound* (1973, Springer-Verlag).
20. M. Morse & U. Ingard, *Theoretical Acoustics* (1968, McGraw-Hill).
21. K. D. Kryter, *The Effects of Noise on Man* (1970, 2nd ed., 1985, Academic Press).
22. L. L. Beranek, ed., *Noise and Vibration Control* (1971, McGraw-Hill). Revised (1988, Inst. Noise Con. Eng.).
23. E. Skudrzyk, *The Foundations of Acoustics* (1971, Springer-Verlag).
24. T. J. Schultz, *Community Noise Rating* (1972, 2nd ed., 1982, Appl. Sci. Pub.).
25a. J. Blauert, *Raumlichen Hören* (1974, Hirzel Verlag).
25b. J. Blauert, *Spatial Hearing* (1983, MIT Press).
26. R. W. B. Stephens, ed., *Acoustics 1974* (1975, Chapman and Hall).
27. R. Mackenzie, ed., *Auditorium Acoustics* (1975, Applied Science Pub.).
28. L. H. Schaudinischky, *Sound Man and Building* (1976, Applied Science Pub.).
29a. L. Cremer & H. A. Müller, *Die Wissenschaftlichen Grundlagen der Raumakustik Vol. 1, Vol. 2* (1978, S. Hirzel Verlag).
29b. T. J. Schultz tr., *Principles and Applications of Room Acoustics Vol. 1, Vol. 2* (1982, Appl. Sci. Pub.).

30. H. Kuttruff, *Room Acoustics* (1st ed., 1973, Elsevier, 4th ed., 2000, Spon Press).
31. V. L. Jordan, *Acoustical Design of Concert Halls and Theatres* (1980, Appl. Sci. Pub.).
32. A. D. Pierce, *Acoustics, an Introduction to its Physical Principles and Applications* (1981, McGraw-Hill) Revised (1989, 1996, Acoust. Soc. Am.).
33. Y. Ando, *Concert Hall Acoustics* (1985, Springer Verlag).
34. A. Lara Saenz & R. W. B. Stephens, ed., *Noise Pollution* [SCOPE 24] (1986, John Wiley).
35. D. Davis & C. Davis, *Sound System Engineering* (2nd ed., 1987, Haward W. Sams, Macmillan).
36. P. M. Nelson, *Transportation Noise Reference Book* (1987, Butterworths).
37. K. U. Ingard, *Fundamentals of Wave and Oscillation* (1988, Cambridge University Press).
38. F. J. Fahy, *Sound Intensity* (1989, Elsevier Science Pub.).
39. Y. W. Lee, *Statistical Theory of Communication* (1960, John Wiley).
40. W. H. Hayt, Jr. & J. E. Kemmerly, *Engineering Circuit Analysis* (1978, McGraw-Hill).
41. F. J. Fahy, *Sound and Structural Vibration* (1985, Academic Press).
42. J. O. Pickles, *An Introduction to the Physiology of Hearing* (2nd ed., 1988, Academic Press).
43. B. C. J. Moore, *An Introduction to the Psychology of Hearing* (3rd ed., 1989, Academic Press).
44. P. A. Nelson & S. J. Elliott, *Active Control of Sound* (1992, Academic Press).
45. K. U. Ingard, *Note on Sound Absorption Technology* (1994, Noise Control Found.).
46. F. Fahy & J. Walker ed: *Fundamentals of Noise and Vibration.* (1998. E & FN Spon).
47. F. Fahy: *Foundations of Engineering Acoustics*, (2001, Academic Press).

Literature C: international and national standards

ANSI S3.5, 1997, American National Standard – Methods for calculation of the speech intelligibility index.

ANSI S12.2, 1995, American National Standard – Criteria for evaluating room noise.

ASTM E 90, 2009, American Society for Testing and Materials – Standard test method for laboratory measurement of airborne sound transmission loss of building partitions and elements.

ASTM E 413, 2004, American Society for Testing and Materials – Classification for rating sound insulation.

ASTM E 492, 2009, American Society for Testing and Materials – Standard test method for laboratory measurement of impact sound transmission through floor-ceiling assemblies using the tapping machine.

IEC 61260:1995, Electro acoustics – Octave-band and fractional-octave-band filters.

IEC 61672-1:2002, Electro acoustics – Sound level meters – Part 1: Specifications.

IEC 60268-16:2003, Sound system equipment – Part 16: Objective rating of speech intelligibility by speech transmission index.

ISO 140-1:1997, Acoustics – Measurement of sound insulation in buildings and of building elements – Part 1: Requirements for laboratory test facilities with suppressed flanking transmission.

ISO 140-1:1997/Amd 1:2004, Specific requirements on the frame of the test opening for lightweight twin leaf partitions.

ISO 140-2:1991, – Part 2: Determination, verification and application of precision data.

ISO 140-3:1995, – Part 3: Laboratory measurements of airborne sound insulation of building elements.

ISO 140-3:1995/Amd 1:2004, Installation guidelines for lightweight twin leaf partitions.

ISO 140-4:1998, – Part 4: Field measurements of airborne sound insulation between rooms.

ISO 140-5:1998, – Part 5: Field measurements of airborne sound insulation of façade elements and façades.

ISO 140-6:1998, – Part 6: Laboratory measurements of impact sound insulation of floors.

ISO 140-7:1998, – Part 7: Field measurements of impact sound insulation of floors.

ISO 140-8:1997, – Part 8: Laboratory measurements of the reduction of transmitted impact noise by floor coverings on a heavyweight standard floor.

ISO 140-10:1991, – Part 10: Laboratory measurement of airborne sound insulation of small building elements.

ISO 140-11:2005, – Part 11: Laboratory measurements of the reduction of transmitted impact sound by floor coverings on lightweight reference floors.

ISO 140-14:2004, – Part 14: Guidelines for special situations in the field.

ISO 140-16:2006, – Part 16: Laboratory measurement of the sound reduction index improvement by additional lining.

ISO 140-18:2006, – Part 18: Laboratory measurement of sound generated by rainfall on building elements.

ISO 226:2003, Acoustics – Normal equal-loudness-level contours.

ISO 266:1997, Acoustics – Preferred frequencies.

ISO 354:2003, Acoustics – Measurement of sound absorption in a reverberation room.

ISO/R 507:1970, Acoustics – Procedure for describing aircraft noise around an airport.

ISO 532:1975, Acoustics – Method for calculating loudness level.

ISO 717-1:1996, Acoustics – Rating of sound insulation in buildings and of building elements – Part 1: Airborne sound insulation.

ISO 717-1:1996/Amd 1:2006, Rounding rules related to single number ratings and single number quantities.

ISO 717-2:1996, – Part 2: Impact sound insulation.

ISO 1683:2008, Acoustics – Preferred reference values for acoustical and vibratory levels.

ISO/R 1996:1971, Acoustics – Assessment of noise with respect to community response.

ISO 1996-1:2003, Acoustics – Description, measurement and assessment of environmental noise – Part 1: Basic quantities and assessment procedures.

ISO 1996-2:2007, – Part 2: Determination of environmental noise levels.

ISO 1999:1990, Acoustics – Determination of occupational noise exposure and estimation of noise-induced hearing impairment.

ISO 2631:1978, Guide for the evaluation of human exposure to whole-body vibration.

ISO 3382-1:2009, Acoustics – Measurement of room acoustic parameters – Part 1: Performance spaces.

ISO 3382-2:2008, – Part 2: Reverberation time in ordinary rooms.

ISO 3740:2000, Acoustics – Determination of sound power levels of noise sources – Guidelines for the use of basic standards.

ISO 3741:1999, Acoustics – Determination of sound power levels of noise sources using sound pressure – Precision methods for reverberation rooms.

ISO 3743-1:1994, Acoustics – Determination of sound power levels of noise sources – Engineering methods for small, movable sources in reverberant fields – Part 1: Comparison method for hard-walled test rooms.

ISO 3743-2:1994, – Part 2: Methods for special reverberation test rooms.

ISO 3744:1994, Acoustics – Determination of sound power levels of noise sources using sound pressure – Engineering method in an essentially free field over a reflecting plane.

ISO 3745:2003, Acoustics – Determination of sound power levels of noise sources using sound pressure – Precision methods for anechoic and hemi-anechoic rooms.

ISO 3746:1995, Acoustics – Determination of sound power levels of noise sources using sound pressure – Survey method using an enveloping measurement surface over a reflecting plane.

ISO 3747:2000, Acoustics – Determination of sound power levels of noise sources using sound pressure – Comparison method in situ.

ISO 3891:1978, Acoustics – Procedure for describing aircraft noise heard on the ground.

ISO 7196:1995, Acoustics – Frequency-weighting characteristic for infrasound measurements.

ISO 8041:2005, Human response to vibration – Measuring instrumentation.

ISO 8297:1994, Acoustics – Determination of sound power levels of multisource industrial plants for evaluation of sound pressure levels in the environment – Engineering method.

ISO 9052-1:1989, Acoustics – Determination of dynamic stiffness – Part 1: Materials used under floating floors in dwellings.

ISO 9053:1991, Acoustics – Materials for acoustical applications – Determination of airflow resistance.

ISO 9612:2009, Acoustics – Determination of occupational noise exposure – Engineering method.

ISO 9613-1:1993, Acoustics – Attenuation of sound during propagation outdoors – Part 1: Calculation of the absorption of sound by the atmosphere.

ISO 9613-2:1996, – Part 2: General method of calculation.

ISO 9614-1:1993, Acoustics – Determination of sound power levels of noise sources using sound intensity – Part 1: Measurement at discrete points.

ISO 9614-2:1996, – Part 2: Measurement by scanning.

ISO 9614-3:2002, – Part 3: Precision method for measurement by scanning.

ISO 10052:2004, Acoustics – Field measurements of airborne and impact sound insulation and of service equipment sound – Survey method.

ISO 10053:1991, Acoustics – Measurement of office screen sound attenuation under specific laboratory conditions.

ISO 10534-1:1996, Acoustics – Determination of sound absorption coefficient and impedance in impedance tubes – Part 1: Method using standing wave ratio.

ISO 10534-2:1998, – Part 2: Transfer-function method.

ISO 10847:1997, Acoustics – In-situ determination of insertion loss of outdoor noise barriers of all types.

ISO 10848-1:2006, Acoustics – Laboratory measurement of the flanking transmission of airborne and impact sound between adjoining rooms – Part 1: Frame document.

ISO 10848-2:2006, – Part 2: Application to light elements when the junction has a small influence.

ISO 10848-3:2006, – Part 3: Application to light elements when the junction has a substantial influence.

ISO 11654:1997, Acoustics – Sound absorbers for use in buildings – Rating of sound absorption.

ISO 11690-1:1996, Acoustics – Recommended practice for the design of low-noise workplaces containing machinery – Part 1: Noise control strategies.

ISO 11690-2:1996, – Part 2: Noise control measures.

ISO/TR 11690-3:1997, – Part 3: Sound propagation and noise prediction in workrooms.

ISO 11821:1997, Acoustics – Measurement of the in situ sound attenuation of a removable screen.

ISO 11957:1996, Acoustics – Determination of sound insulation performance of cabins – Laboratory and in situ measurements.

ISO 13474:2009, Acoustics – Framework for calculating a distribution of sound exposure levels for impulsive sound events for the purposes of environmental noise assessment.

ISO 14257:2001, Acoustics – Measurement and parametric description of spatial sound distribution curves in workrooms for evaluation of their acoustical performance.

ISO 15186-1:2000, Acoustics – Measurement of sound insulation in buildings and of building elements using sound intensity – Part 1: Laboratory measurements.

ISO 15186-2:2003, – Part 2: Field measurements.

ISO 15186-3:2002, – Part 3: Laboratory measurements at low frequencies.

ISO 15664:2001, Acoustics – Noise control design procedures for open plant.

ISO/TS 15666:2003, Acoustics – Assessment of noise annoyance by means of social and socio-acoustic surveys.

ISO 15712-1:2005, Building, Acoustics – Estimation of acoustic performance of buildings from the performance of elements – Part 1: Airborne sound insulation between rooms.

ISO 15712-2:2005, – Part 2: Impact sound insulation between rooms.

ISO 15712-3:2005, – Part 3: Airborne sound insulation against outdoor sound.

ISO 15712-4:2005, – Part 4: Transmission of indoor sound to the outside.

ISO 16032:2004, Acoustics – Measurement of sound pressure level from service equipment in buildings – Engineering method.

ISO 17497-1:2004, Acoustics – Sound-scattering properties of surfaces – Part 1: Measurement of the random-incidence scattering coefficient in a reverberation room.

ISO 17624:2004, Acoustics – Guidelines for noise control in offices and workrooms by means of acoustical screens.

ISO 18233:2006, Acoustics – Application of new measurement methods in building and room, Acoustics.

ISO/TR 25417:2007, Acoustics – Definitions of basic quantities and terms.

JIS A 1418-1, Japanese Industrial Standards – Measurement of floor impact sound insulation of buildings; Part 1: Method using standard light impact source.

JIS A 1418-2, – Part 2: Method using standard heavy impact source.

JIS A 1419-2, Japanese Industrial Standards – Rating of sound insulation in buildings and of building elements; Part 2: Floor impact sound insulation.

JIS A 1440, Japanese Industrial Standards – Laboratory measurement of the reduction of transmitted tapping machine impact sound by floor coverings on a solid standard floor.

Index